Expect More:
Children Can Do Remarkable Things

Anne Grall Reichel, Ed.D.

With **Becky Gill, Ed.D. & Joanne Efantis Trahanas**

authorHOUSE®

AuthorHouse™
1663 Liberty Drive
Bloomington, IN 47403
www.authorhouse.com
Phone: 1-800-839-8640

First published by AuthorHouse 9/16/2010

ISBN: 978-1-4520-7632-4 (e)
ISBN: 978-1-4520-7631-7 (sc)

Library of Congress Control Number: 2010913423

Printed in the United States of America

This book is printed on acid-free paper.

For my husband, Jack
My constant

Table of Contents

Acknowledgements

I have had the privilege of working with Dr. Becky Gill for at least ten years. Becky is the Director of Curriculum for Barrington School District 220 in Barrington, IL. Becky leads with passion and intelligence. The work we have accomplished together in Barrington would not have been possible without her eye for detail and her ability to always set high expectations for herself and others. I consider Becky to be both a friend and colleague. I am eternally thankful that she is part of my life. Her insights throughout the writing process were inspirational.

I have had the privilege of working with Joanne Efantis Trahanas for the last three years. Joanne is Curriculum Director for River Forest Elementary School District in River Forest, IL. Joanne's love for learning is evident in her approach to the professional development of her teachers and the curriculum leadership she provides. I am most appreciative of the insights and suggestions she provided throughout this writing process.

I would like to acknowledge the Oak Park/River Forest Science Articulation Team made up of teachers from Oak Park Elementary District, River Forest Elementary District and Oak Park River Forest High School. The continuums were an outgrowth of their expressed interest in clarification of expectations for teaching science. Their review of the continuums created wonderful discussion and perspective. I have deep admiration for their professionalism.

I would like to thank the many teachers in Barrington District 220 who have attended curriculum development meetings and professional development throughout the years. Their professionalism and wonderful questions have fueled my own learning process.

I would like to acknowledge Barry Bertsch of Barry Bertsch Photography for the photograph on the back cover.

Invitation

Several months ago I came across a Dave Granlund cartoon that captured my attention. Granlund cleverly illustrated a museum corridor with the signage **"Extinct Species"**. One display case contained a dinosaur skeleton, another the remains of a prehistoric fish, the third a young man dressed in winter attire holding on to a snow shovel. Now I don't want you to think that my intent is to promote the development of snow shoveling kids. My intent is to invite you to consider the entrepreneurial spirit and intellect of the snow shoveling kid.

Let's consider the successful snow-shoveling entrepreneur. First of all, to run a successful snow shoveling business you had to be a self starter, you had to be personable enough to generate customers, innovative enough to create your own marketing campaign, and savvy enough to carefully watch the weather patterns to prepare for business. You had to have enough business sense to figure out your start up costs and perhaps find a way to finance them. On the morning of the snowstorm when all the other kids were nestling under blankets watching TV you had to have the drive and persistence to brave the cold and follow through on your commitment.

The young snow-shoveling entrepreneur speaks to me on many levels. My entrepreneur could best be described as a persistent, integrative thinker with exceptional follow through skills. In our complex post-modern world of immediacy and instant gratification the perseverance exhibited by my entrepreneur seems to be a lost art. Reflective, integrative thinking, careful planning, high expectations, and problem solving are at best on the endangered list.

This book is an invitation to teachers and parents to embrace the notion of regenerating the entrepreneurial spirit in our children. By "entrepreneurial spirit" I mean having that willingness and perseverance to embrace challenges, the interest in solving problems, and the confidence to take risks. Simply stated, we need to expect more. In my work with teachers, and occasionally parents, it has become increasingly clear to me that we do not have consistently high academic expectations for all children. We expect more from some than others and all too frequently the challenge level of what we expect does not increase with nearly enough complexity as students move through the grades. Our children are continually programmed to move from one after school activity to the next with little time for cognitive rest. Children have little time for the kind of unstructured play that allows imagination and creativity to flourish. In *Last Child In the Woods*, author Richard Louv eloquently discusses the fact that our children have moved indoors and video games have become their imagination.

Children are in dire need of challenging curriculum that allows them to explore complex ideas and opportunities to apply those ideas in new situations. They need opportunities to create and to imagine new ways of doing things. The US Department of Education reported that in 2007-2008 approximately a third of first year college students were in need of at least one remedial course.

I am concerned that somewhere along the way we stopped challenging children and started enabling them instead. Perhaps this is through no fault of our own? As teachers and parents we genuinely care about children. We feel for them when they are sick, when they are bullied or when they are struggling. Quite simply it is our nature to protect. With the best of intentions we tend to try and eliminate the struggle. But the struggle is essential to growth and a personal sense of accomplishment. As children work their way through challenges they build the confidence and habits of mind needed to embrace the next challenge that comes along.

I agree full heartedly with those who suggest that we need to increase the academic challenge for all students. New Common Core Standards recommend the promotion of critical thinking, problem solving, and the development of clear communication skills across the disciplines. The Common Core standards recommend that students spend more time reading challenging informational text and more time supporting their "claims" with "evidence". The new Common Core Standards demand that we raise the bar and expect continuous growth from all students. To demand anything less will surely lead to our demise in a highly innovative competitive global market.

The skills that will be required of our children include "systems-thinking", the ability to support "claims" with "evidence", the ability to communicate to argue, inform and narrate, and the ability to make informed decisions based upon a strong foundational understanding of both science and social studies. Unfortunately science and social studies have barely simmered on the back burner and still remain the stepchildren of many elementary curricula.

My goal, in writing this book, is to develop something useful to both teachers and parents in their work with children. My work promotes increasing the complexity of what we challenge children to do while promoting clarity in the expectations we set forth. The ultimate goal is to nurture a generation that can make informed decisions based upon a strong foundational understanding of science and social studies.

The work of educating children who can innovate and compete in an increasingly competitive global environment cannot be placed solely in the hand of the school. Responsibility for educating children must also rest with the parents, grandparents, volunteers who work with children, neighbors and friends. Yes, it will take a village! I am hopeful that you will read on and embrace the challenge.

All too often, I find myself alarmed by the evidence I gather of the exponential rate at which we choose to "dumb down" America. Once a previous neighbor apologized for referring to my home as the "library". I told her there was no need to apologize and that actually I was quite flattered by her reference to my home. That resulted in a look of sheer puzzlement. Somewhere along the way the prevailing American theme became one of referring to those who read as "bookish," those who tinkered as "geeks," and those who grew passionate about a subject as "unusual".

The example above was not in isolation. Over and over, when I have had the opportunity to end up at the health club for my swim when young mothers are in the locker room I have heard the complaining about crayons and other enemies that invade homes when children are on vacation. I've heard young mothers discuss their concerns about the child who would rather spend the day reading or refer to their child who would rather solve math problems or dig under a rock as the "unusual one" in the family. My advice is really quite simple, embrace the crayons, embrace the questions, embrace the digging under rocks and please take time to put down the cell phone and converse with the kid who chooses to solve math problems or reads lots of books. These are the kids who might lead us if we stop making them feel "unusual" because they are passionate about learning or simply curious. We keep "dumbing down" the thinkers, explorers, and readers because over and over – even without meaning to do so, they hear us say they are "unusual". They learn that to "fit in" they should be less passionate about learning. Please do not simply celebrate their passion for inquiry, they already know they are passionate. Nurture their passion by asking questions, providing resources and genuinely taking time to learn along with them. As children read and problem solve they build the skills that are ultimately performed in the disciplines of science and social studies.

We are in dire need of thinkers, of conversers, and of children who embrace the habits of mind of scientists and social scientists. We can do it if we simply embrace the challenge. It will require prioritizing time with children, getting children outdoors to explore, getting children to the library to gather books on a topic of interest or to the appropriate internet resources, and taking the time to embrace their interest. Most importantly it will require our own passion for learning something new.

I invite every teacher, parent, grandparent, volunteer, neighbor and friend to take on the challenge of supporting a child in learning something new and insisting that for each thing that child learns that they are encouraged to support their ideas with on-going evidence and explanation. We can do it. Children desperately need to see adults model their own passion for learning new things. When we pick up a book or go to the internet to research something we are not certain about, we model the importance of learning for a lifetime. It might require a shift in our priorities, but I guarantee that any adult who embraces the challenge has nothing to lose and a great deal to gain.

Organization & Focus of the Book

Each chapter of this book will be organized in the same way. I will begin with a brief **research base**. I want to be certain that as a reader you understand that the intent of this book is not to provide the research, others have done so. My intent is to reference the research so you can investigate further if you are interested in doing so. It simply will legitimize my approach.

After a brief introduction to the research base in each chapter you will find a **continuum** broken into grade level bands representing the chapter topic. For example, in Chapter One I will explore the development of "Systems Thinking". My grade level bands follow the work of the American Association for the Advancement of Science in *Benchmarks for Science Literacy* clustering development of skills and concepts into K-2, 3-5, 6-8 and 9-12 bands. Just as easily as using grade level bands one could apply progressive language to development such as Bonnie Campbell Hill did in developing literacy continuums in 2001. Words such as emerging, developing, bridging and fluent along with a myriad of others have been used as descriptors since the emergence of continuums in language arts. The important thing to recognize is that not all children develop at the same rate.

As adults, embracing the challenge of developing thinkers, our work is to simply take children from where they are and bring them to the next level so they can experience success and most importantly the joy of learning and of accomplishment. Instead of beating children down by telling them they did not reach "The Score" lets build them up by helping them reflect on what they have accomplished and where they need to go next! Doing this will be what helps them reach "The Score" with out them even knowing it.

In my mind this is a much more positive approach to education. After all, I am willing to bet that there is not a reader of my book who cannot relate to Freud's observation that the confidence experienced in success leads to further success. It is our job to challenge so that children can experience the joy of the pursuit.

The continuums I have developed focus on applying the research to reflective tools in "kid friendly" language so that children can take charge of their own learning. Adults can use these continuums to help children focus on the details of their work. The two subsequent sections in each chapter will focus on practical suggestions for **teachers** and **parents.** I will make suggestions as to how both teachers and parents can work with my tools for science and social studies. The suggestions provided for teachers and parents, can readily be used by other adults committed to furthering the development of thinkers and problem solvers. Volunteers, working in nature-based and informal site work with children, will find the scientific drawing, graphing, systems thinking and communication continuums to be helpful in informing their expectations and feedback for children.

Background: Evolution of My Continuums

In my work as a curriculum consultant, I have often observed that teachers in different grade level bands rarely have an opportunity to articulate a common vision for how skills develop and progress from grade to grade. Elementary, middle school and high school teachers often function as separate entities. During the 2009-2010 school year I had the opportunity to work with the Oak Park/River Forest Science Articulation Team. The teachers expressed an interest in developing a common vision for science skills. During the same time as I was working with the consortium the Common Core Standards for English Language Arts & Literacy in History/Social Studies, Science, and Technical Subjects were published. I was thrilled to see the integration of reading and writing in the content areas. At grades 6-12 the

Writing Standards for Literacy in History/Social Studies, Science and Technical Subjects are clearly defined. At grades K–5 the writing standards are written to cross all subject areas. I decided to take a proactive stance and develop continuums in "kid friendly", reflective language so that teachers could readily see the progression of writing in science throughout the grades. The continuums were informed by the language within the Common Core Standards and were reviewed by the Oak Park/River Forest Articulation Team. I appreciate the thoughtful insights of this group.

In the Common Core Standards at Grades K–5 the language "claims" and "evidence" is not used. I added this language to the writing continuums because it is the language of science. I have observed first graders in a Chicago urban school readily embrace this language when provided with appropriate scaffolding and modeling. The continuums are written in grade level bands with the expectation that students would hit these targets by the end of the grade level band. The continuums are written in reflective language so students can take charge of their own writing and thinking. As mentioned above, do not let the grade level bands limit your work, use them to access where children are in their writing and thinking and help them to move to the next place. Our charge is to nurture growth!

Chapter One: Systems Thinking

"All great entrepreneurs are systems thinkers"
Michael Gerber

From futurists to business leaders, there is no shortage of individuals who have heralded the importance of "systems thinking". David Thornburg identified "systems thinking" as one of the essential skills of the 21st Century. Peter Senge brought "systems thinking" to the forefront of the business world. In *A Whole New Mind,* Daniel Pink pointed out that both Charles Schwab and Richard Branson credit their success with being able to see the "big picture".

As early as 1990, The American Association for the Advancement of Science (AAAS) highlighted the importance of systems thinking in their highly celebrated book *Science for All Americans.* In 1993 when AAAS developed *Benchmarks for Science Literacy* they made the case that systems thinking was an essential component of higher-order thinking. They stated; "One of the essential components of higher-order thinking is the ability to think about the whole in terms of parts and, alternatively, about parts in terms of how they relate to one another and to the whole. ...The scientific idea of systems implies detailed attention to inputs and outputs and to interactions among the system components." (AAAS, p.262)

By now you would probably like a clear explanation of "systems thinking". Simply put, systems thinking means having the capacity to understand how things within a system influence one another. Systems thinkers are capable of seeing patterns and relationships within systems. Rather than seeing events in isolation they view them in relationship to the whole. As teachers and parents we use systems thinking in understanding the dynamic relationships within our classrooms and families. We recognize patterns that we have seen in the past and are capable of using those patterns and previous experiences to keep our operations running smoothly. We recognize that single actions or events do not occur in isolation. As classroom teachers we work with many dynamic systems that help our classrooms function effectively. We recognize that from our systems of management to our systems for analyzing data and differentiating instruction, that all systems are inextricably connected to one another. We constantly have an eye on the big picture and adjust often to keep our classroom systems functioning.

In *Benchmarks for Science Literacy*, AAAS did a beautiful job of describing how "systems thinking" progresses throughout the grades. The AAAS benchmarks, and insights from the National Science Education Standards were used to develop the "Systems Thinking" continuum below. All continuums are written in "student friendly" reflective language, in order to put students in charge of their own learning process. After developing the Systems Thinking continuum, it was reviewed by the Oak Park/River Forest teachers who encouraged me to provide examples so teachers and parents could readily think about some tangible examples. While the examples below the continuum relate to science, please note that in both science and social studies there are many opportunities to apply "systems thinking". In social studies systems abound from political to economic systems. One cannot effectively investigate cause/effect relationships without applying systems thinking. Below the continuum you will find specific suggestions for both teachers and parents and a brief discussion of the relationship between systems thinking and pattern recognition. Patterns of change will be further developed in the chapter on graphing.

In conclusion, if we are committed to developing critical thinkers and problem solvers, then we are compelled to see the importance in developing systems thinking. The continuum on systems thinking provides clarity and focus for talking about systems with children. The continuum helps us see how the concept of systems thinking develops over time and provides teachers and parents with a progression of expectations they can use when facilitating learning experiences. It is exciting to think about the dynamic conversations and potential for student engagement if we embrace the challenge of teaching systems thinking.

Systems Thinking – Continuum Grades K-12 - © 2010 (Anne Grall Reichel) All Rights Reserved

K-2	3-5	6-8	9-12
I identified the parts of a system.	I identified the parts of a system.	I identified and described the relationship between parts of a system.	I identified and described the relationships between parts of a system.
I predicted what would happen if part of the system was missing.	I described the "job" (function) of a system "part" (structure).	I related structures to functions.	I recognized that complex systems may have components to detect, backup, bypass or compensate for minor failures.
I tested my prediction.	I described how the parts of a system affect one another.	I made predictions about a system's ability to function if parts of the system are missing or broken.	I measured input and output to a system and described ways to make the system more efficient.
I described my test results.	I recognized that a system may not work as well if a part is missing, broken, worn out or misconnected.	I identified system input and output.	I tested the efficiency of a system.
	I inspected a system that is not working and identified what to do to make it work.	I recognized that output from one part of a system can be input to another part of the system.	I analyzed a system and made predictions about how input and output will impact the system.
	I made changes to a system to correct a problem.	I identified the necessary changes to make to a system based upon my investigation of system input and output.	I tested predictions and made suggestions based upon my results.
	I understand that the solution to one problem may cause new problems.	I identified a feedback loop within a system.	I identified layers of control within complex systems.
		I made predictions about how output that is fed back into the system will impact the system over time.	I recognize that even complex automatic systems with layers of control require human intervention at some point.
		I recognize that mechanical systems are sometimes quicker but still require human oversight.	I applied systems thinking to trouble-shoot system failures.
		I used systems thinking to trouble-shoot system failures.	I discussed the advantages and disadvantages to problem solutions recognizing that solutions may be at the root of new problems.
		I recognize that solutions may lead to new problems.	I recognize that many new systems will be needed to respond to the complexities of life in the 21st Century.

Systems Thinking – Examples Grades K-12

Examples:

- Rearrange parts of simple objects (gear toys, dolls, blocks, etc.,)
- Find out what happens if a plant part is missing.
- Explore the role of body parts, predict what would happen if a part were missing.
- Describe parts and their jobs: (i.e. jointed insect legs for jumping, wings for flying, etc.,

Examples:

- Build rubber-band and balloon powered cars.
- Investigate how a change in wheel size or composition affects the vehicle.
- Investigate "trade offs" when using simple machines.
- Troubleshoot simple parallel and series circuits.
- Investigate roles within food webs.
- Build simple eco-columns and observe and alter parts over time.
- Predict what would happen if there was a sudden change in one population within an ecosystem.
- Investigate the relationship of body parts: (i.e. The brain gets signals from body parts, the brain sends signals to body parts).
- Investigate water flow and the impact on weathering and erosion.

Examples:

- Build mouse trap cars and manipulate parts evaluating the overall success of the system.
- Use graphing calculators to manipulate movement of small vehicles.
- Explore connections among systems.
- Compare and contrast navigation with simple and complex systems: (i.e. compass and GPS)
- Investigate system input and output: (i.e. Photosynthesis CO_2/O_2)
- Compare and contrast open and closed systems: (i.e. evaporation in closed and open systems)
- Examine levels of complexity in systems: (i.e. cell, tissue, organ, organ system – particles, atoms, molecules)
- Examine simple feedback systems: (i.e. furnace/thermostat)
- Determine efficiency of systems (i.e. solar panel/longevity)
- Investigate cases of human error in operating technological systems: (i.e. Metric Mishap/Mars Orbiter)

Examples:

- Investigate the role of electrochemical balance in relationship to the nervous system.
- Investigate the interactions between the immune and circulatory systems.
- Investigate health of the human system in relationship to our choices.
- Compare and contrast open and closed systems.
- Investigate the role of a catalyst in a chemical reaction.
- Investigate changes in a system in relationship to equilibrium. (i.e. Small disturbances may settle to the same state of equilibrium, large disturbances may lead to escape from that state and the settling into a new state of equilibrium.)
- Investigate changes in systems in systems as evidence of evolutionary change.
- Investigate how systems can change in detail but remain the same in general: (i.e. cells are replaced, but the organism remains.)
- Examine systems and their limitations: (i.e. computers, GPS systems, etc.,

Please Note: Examples come from a variety of sources including science curriculum, the National Science Education Standards, and Benchmarks for Science Literacy

FOR TEACHERS

IMPLEMENTING CONTINUUMS:

As you begin to think about implementing continuums in your classroom my first suggestion is to focus on your grade level band.

If you are a K-2 teacher use the next level of the continuum to differentiate for children who need a greater challenge. If you teach third grade or above, use the other grade bands to consider ways to differentiate for students who either need a greater scaffold or a greater challenge.

CREATE CLASSROOM CHARTS

Creating a chart of the continuum expectations for your grade level band creates a focal point for instruction and student reflection. Barb Mayer, a second grade teacher in River Forest, was one of the first to begin piloting my continuums in her classroom. Barb shared that as she worked with the reflection points in a single continuum she made a large class chart that was posted in the classroom for her grade level band of the continuum. She found that she could refer to the chart and challenge her children to reflect on their work. In Barb's words using the continuum chart was helpful because, "children knew just what to do". The charts make the correct vocabulary visible to us as facilitators of student learning, they give the student the correct vocabulary for what they are doing. The literacy experts have helped us understand the importance of developing common vocabulary that is used at both home and at school. Continuums provide the means to do so. A sample classroom chart for Systems Thinking in grades K–2 appears below.

Systems Thinking

I identified the parts of a system.

I predicted what would happen if part of the system was missing.

I tested my prediction.

I described my test results.

As I reflected on Barb's comment I realized that all too often we assign projects without making our expectations transparent or without making the expectations and the accompanying language major components of the lessons as well as the resulting work. As teachers it is critical that we model our own use of the continuum and point out exactly how each component plays out in the work samples we provide. Of course, the best samples are authentic student work samples generated over time.

5

Anne Grall Reichel, Ed.D.

USE CONTINUUMS TO FOCUS FORMATIVE ASSESSMENT

The continuums can be used to formatively assess student progress and to help students focus in on exactly what they need to do next. For example, primary children working with a simple system such as a plant may be challenged to identify essential components of the plant and the simple functions of each part. Students can reflect on their own drawings by checking the chart and then seeing if they have "identified all the parts of the system".

Let's continue to play out the K–2 continuum above. Teachers can go on to formatively assess student understanding of a "fair test" by moving on to the remainder of the continuum. "Fair Test" is terminology we use in the primary grades as a means of working towards an understanding of a controlled scientific investigation in later grades, only one of the methods a scientist might use. A fair test simply means we keep everything in the experiment the "same" with the exception of the one thing we are testing. For example, if we were interested in determining if temperature impacts seed germination the only factor we would change would be temperature. We would keep all other conditions the same. We would use the same number of seeds, the same kind of seeds, the same amount of water, the same sized container and we would make certain the lighting conditions were the same.

Working towards an understanding of systems thinking children can be challenged to design a "fair test" to determine what would happen if a critical plant part were missing. They can make a prediction based upon their understanding of plant parts and can test their prediction by starting with two plants, removing the part from one and not from the other. The plants can be kept in the same conditions. Preliminary data can be gained based upon the one experiment the child performs. This is called preliminary data because, of course, we would communicate that scientists would not draw sweeping conclusions based upon one isolated result. We should always encourage our young scientists to continue their own testing and to ask other scientists, their classmates, to verify their findings. Encourage the children to state a "claim" and provide the supporting "evidence". This skill, constructed through hands–on experience in science, rests at the backbone of constructing an intelligent argument or persuasive essay. It is also a lovely way to generate a community of scientists in your own classroom. It provides a forum for children to talk about important things and raises the bar on the level of the conversation. I have heard primary students challenge one another to provide the evidence for their claims. Just think of the amazing thinkers and writers we can develop if we begin with concrete experiences like the one above in the primary grades. We can expect more and we can get more as a result!

Continuum Descriptor	What Students "Do" and "Say"	Point for Formative Assessment
I predicted what would happen if part of the system was missing.	I predicted that if I took the flowers off the plant that the plant would not make new seeds.	Did the student make a logical prediction based upon his or her understanding of plant parts?
I tested my prediction.	I removed the flowers from my fast plants just as soon as they started to bloom.	Can the student design a "fair test" changing only one variable at a time?

The continuum provides key points for formatively assessing. As you look at the progressive levels of the continuum you will observe that complexity of systems thinking continues with third through fifth graders inspecting and modifying systems and sixth through eight graders going on to consider inputs and outputs. The point here is simple, move beyond simple systems as you progress through the grades. All too often I will observe assessments where the "systems thinking" does not progress incrementally, simply stated continue to raise the bar guided by the continuum. Expect more!

ORGANIZING CURRICULUM AROUND "BIG IDEAS" LIKE SYSTEMS

Over a decade ago I began supporting school districts in developing science and social studies curriculum around "big ideas". The "big ideas" are sometimes referred to as unifying concepts. They include ideas like systems, patterns, change and constancy and scale. I am not the first to recommend the use of "big ideas" to organize curriculum. Heidi Hayes-Jacobs was one of the first leaders in this field. What teachers tell me is unique in my approach is the emphasis on the growth of these unifying concepts from year to year. For example, if a kindergarten science curriculum is organized around "systems of sorting" the big idea is not abandoned after kindergarten. Students continue to broaden their schema for "systems of sorting" as they progress through the grades. A young child may sort by color, texture, shape, sink/float, magnetic/non-magnetic, a fourth grader may sort materials as conductors and insulators, an eighth grader might construct an understanding that the Periodic Table of Elements is a sophisticated system of sorting. The idea is to apply the concept in multiple contexts throughout a single grade level and beyond. In "How Students Learn" the National Research Council emphasizes the importance of teaching concepts in multiple contexts. In both, Barrington Schools and River Forest Schools elementary teachers use "big ideas" to organize science curriculum. A visualization of a sample K–5 progression appears below.

Unifying Concepts
A new concept is introduced in each progressive grade
Previous concepts are revisited in new contexts

					Systems & Interactions
				Change & Constancy	Change & Constancy
			Systems & Relationships	Systems & Relationships	Systems & Relationships
		Patterns of Change	Patterns of Change	Patterns of Change	Patterns of Change
	Change Over Time	Change Over Time	Change Over Time	Change Over Time	Change Over Time
Systems of Sorting	Systems of Sorting	Systems of Sorting	Systems of Sorting	Systems of Sorting	Systems of Sorting
Kindergarten	**First Grade**	**Second Grade**	**Third Grade**	**Fourth Grade**	**Fifth Grade**

Please see Appendix A for an example of unifying concepts played out across the grades.

Using the chart above let's consider how systems thinking increases in complexity as students move through the grades. By third grade "systems thinking" grows as students are challenged to consider systems and relationships. Third graders investigating systems and relationships might investigate plants as systems. They extend their study of a plant as a simple system to plants within a prairie ecosystem. The third graders explore the relationship between a plant and a pollinator in life science. In physical science they might explore the relationship between the mass of an object and the amount of force needed to move the object a distance or they might investigate the "trade off" of distance to force as they work with simple machines. In earth science third graders might investigate the relationship between the position of earth, moon and sun and the phase of the moon we observe from earth. This affords the children with the opportunity to construct an understanding of the concept that systems do not exist in isolation. **Students begin to recognize that intricate living systems depend on other systems to carry out processes**. If you follow the third grade progression you will see that third graders go on to investigate concepts of force and motion in physical science and in astronomy they are introduced to the earth, sun, moon system. Through a great deal of hands-on simulations and modeling they begin to construct a nascent understanding of the notion that moon phases and other sky phenomenon are dependent upon earth's relationship with other sky objects.

By fifth grade students can extend "systems thinking" to interactions. They can investigate the human body as composed of an intricate series of systems that interact. In physical science they can be challenged to consider interactions as they further their understanding of force and motion. For example, a simple inquiry might look at what happens when a moving object interacts with another object. Students might go on to investigate the interactions along a river system between water and land in earth science. The potential is limitless. The idea is to build the complexity of systems thinking by using the continuum levels as a guide while increasing the understanding of fundamental science concepts.

MAKING CONNECTIONS

By using unifying concepts or big ideas in our teaching we can support children in making connections. Marilyn Ferguson, a renown psychologist and author stated; "Making mental connections is our most crucial learning tool, the essence of human intelligence; to forge links; to go beyond the given; to see patterns, relationships, context." The "systems thinkers" highlighted in the chapter introduction all attributed their success to recognizing the patterns and relationships within the big picture. In *How People Learn* (2000) Bransford, Brown & Cocking highlight DeGroot's (1965) research comparing strategies used by expert and novice chess players. DeGroot observed that expert chess players could consistently out-think their opponents. He found that the expert chess players were adept at perceiving patterns in chess configurations that escaped the novice players. The expert chess players strategically used the patterns to formulate their next move. Developing children's ability to recognize patterns sets up the optimum conditions for them to access the knowledge that is needed to complete whatever task is at hand. It is now well documented in the research base that pattern recognition is an important strategy in supporting students in developing both "confidence and competence" (Bransford, Brown & Cocking, 2000, p. 48).

ANCHOR CHARTS

As teachers, it is critical that we make time in our day for children to participate in conversations that are focused on key concepts. This allows children to participate in meaningful discussion and to build connections. I've discovered that anchor charts serve as a very helpful tool in keeping an on-going record of the big ideas and connections that children make. More importantly, I believe the anchor chart has phenomenal potential to serve as a tool for further inquiry. By setting up a systems chart we can promote an atmosphere of inquiry by encouraging children to look for other examples of systems. We can have important conversations about "systems" and use our Anchor Chart to generate conversation on how systems are alike and different. Miller (2002) affirms that anchor charts make thinking visible. "Having previous ideas visible helps children make connections and think more deeply about their experiences and how they are related to the unifying concept they are studying" (Sahn & Reichel, 2008, p.15). A sample Anchor Chart appears below.

Systems at School	Systems at Home	Systems in the Community
Plant Animal Number System Letter System System to "Write to Argue" System to "Write to Inform" Attendance system Heating system Alarm system Computer	Appliance as system (refrigerator, oven, dishwasher, coffee pot) System for getting dinner on the table System for communicating with family members System for getting homework done Pet as system dependent on other systems	Fire Department Police Department Transportation Systems Recreation System Pond River Prairie Park

FOR PARENTS

As I begin this first section for parents, I must admit that I think the challenges of parenthood are greater than ever. We are moving at a frenetic pace, we all seem to be on fast forward. My first piece of advice is this; If you are interested in nurturing the growth and development of a "systems thinker", then slow down and model your thinking. By this I simply mean when you have to solve a problem or engage in a house or yard project that requires systems thinking make your thinking transparent as you go. Here are a few vignettes to help you visualize the idea.

TROUBLE-SHOOTING APPLIANCE FAILURE

Let's say the coffee pot is not working. Many times the solution is simple, but it requires "systems thinking". Are all the parts connected correctly so that there is contact between parts? Is the coffee pot plugged in? Is the switch in the right position? Are the correct buttons pushed? Is there an indicator light telling you something is out of synch? We are all programmed to just quickly go down that list or worse yet we throw up our hands and ask someone else to fix it rather than engaging in the process ourselves. The point is, if children never hear or see you think through simple problems, they will have no idea how to solve them. If they don't see you embrace challenge they will be disinterested in challenge.

PLANTING FLOWERS

Let's say you want to plant summer flowers. Before you go to purchase your flowers take the time to go outside with your child and inspect the place where you would like to plant. Explain that first you need to check out if it is a sunny or shady spot. Encourage your child to figure out a way to determine whether the spot is sunny or shady. Give them a few days to figure it out. Next take your child along to purchase the flowers. Don't just go grab the flowers. Based upon the determination of "sunny" or "shady" ask your child to figure out which plants to purchase. Model your own inquiry as you read the specific needs of the plants on the labels. On the way home discuss those with your child and come up with a plan. Implement the plan together and give your child responsibility for the care. Expect more!

I know, by now you are probably saying; "Are you kidding me, who has time for this?" My advice, make time. One of my favorite pass times at traffic lights is to observe the interactions of children and parents when they are stopped next to me. All too often Mom or Dad is on the cell phone and the kids are clicked into something digitized in the back seat. I see this phenomenon in grocery stores, restaurants and parks as well. We have become addicted to our phones and our children are becoming more and more detached. A colleague told me a story about her 30 something daughter who came to her very upset. Her daughter shared that she was reading to her 3 year old one evening when the three year old stated; "Excuse me Mommy I have to take this call." The child proceeded to pretend she was taking a call. The story speaks volumes. Our children model what they see. If we want thinkers and problem solvers then we have to model what that looks like. We have to engage in solving problems with them.

LEAVE NO CHILD INSIDE

Currently there is a national movement focused on reconnecting children with nature. The movement was largely inspired by Richard Louv's book *Last Child in the Woods*. Louv coined the term "nature-deficit disorder" referring to "the disconnect" between children and nature. Nature is the perfect place to engage in "systems thinking". It might be as simple as engaging your child with a simple question. "How could we encourage more birds to visit our backyard?" (National Wildlife Foundation has wonderful information on their website to support questions such as the one above.) The point here is model your thinking as you and your child figure this out. Projects such as this are not limited to suburban families. Urban families can explore similar inquiries with "Project Pigeon Watch". Information is readily available on the internet.(Please see Appendix B for a listing of links and descriptions)

Author's Notes:

Over the past few months I have marveled at the lack of "systems thinking" that is evidenced when discussing the Gulf Coast Disaster in 2010. For the first month of the spill I watched the news in disbelief. My husband and I spent dinner conversations talking about it. We concluded that the average American would not be concerned until the oil hit the coast and images of the impact on wildlife emerged. That is exactly what happened. One day I overheard a woman in her twenties remark that she really didn't think about it until she went to a seafood restaurant. Such remarks speak to the scientific literacy of the average American and the lack of systems thinking evident in everyday interactions. For example, if you are a Sushi lover who enjoys blue fin tuna that are caught off the New England Coast, without the use of systems thinking you might be inclined to think your sushi choices are safe. Think again! Blue fin tuna caught off the coast of New England are remarkable systems in and of themselves. Those blue fin tuna take a remarkable swim before they are caught and expeditiously consumed by the Sushi lover. Some blue fin tuna spawns in the Gulf of Mexico. If we employ systems thinking we quickly start to make the connections. See this is it! We can do much better, systems thinking allows us to see our interconnections.

HOW CAN PARENTS USE THE CONTINUUM?

Parents can use the continuum as a starting point for coming up with questions that promote systems thinking. Take a look at the following chart. I used the Grade 3-5 band to create an example for you.

Continuum Descriptor	Parent Question
I identified the parts of a system.	What are the parts of this system?
I described the "job" (function) of a system "part" (structure).	What do you think this part does?
I described how the parts of a system affect one another.	How do you think this part works with other parts?
I recognized that a system may not work as well if a part is missing, broken, worn out or misconnected.	What if this part was missing? If this part was missing how could you fix it?
I inspected a system that is not working and identified what to do to make it work.	Can you figure out how to make this work?
I made changes to a system to correct a problem.	What changes did you make?
I understand that the solution to one problem may cause new problems.	Did your solution create any new problems?

Questions, such as the ones in the chart above can be applied to any system. The questions lead to interesting discussion. As parents remember the significance of questions. I credit questioning to the most significant turning point in my own teaching career. One year I decided to monitor my own ability to respond to the questions that my 8th graders asked me in science class with new questions. Simply stated, I started asking questions of questions.

The inquiry and rich conversation increased exponentially. So as a parent the next time your child asks a question ask one back instead of supplying the answer. This is a great way to encourage thinking and to engage in meaningful conversation.

ABOUT CONVERSATION

Before concluding this chapter I would like to briefly reflect on the role of conversation in developing systems thinkers. My mother and father encouraged systems thinking every night at the dinner table. At some point I figured out that every night before dinner my father either read something out of the newspaper or most frequently opened the World Book Encyclopedia to a random page. He then cleverly turned whatever he read into a debatable question that ultimately required one member of the family leaving the table and fetching the correct volume of World Book for further discussion. Of course that often led to gathering further evidence by linking to other volumes. The conversations were lively and interesting. Dinner often took a long time. To this day I think there was something metaphorical in the fact that we sat at a round dinner table. As a child, I really traveled around the world at the dinner table and experienced the joy of pursuing knowledge and understanding. My parents nurtured life long learning skills. So yes, I am advocating turning off the television and the cell phone during dinner. I am advocating family conversation around intelligent meaningful subjects, ones that require systems thinking to solve them. Trust me it can be really fun.

Chapter Two: Technological Design

Technological Design

"The problems that exist in the world today cannot be solved
by the level of thinking that created them."
Albert Einstein

The charge to teach "technological design" emerged in The National Science Education Standards in 1996. To this day, I still remember my own fascination with the process described in this standard. It seemed to me that the notion of promoting a thinking process that would engage students in looking at problems, proposing solutions, designing, building, testing, evaluating and modifying held great promise. I recognized that the skills of technological design required perseverance and systems thinking. While I knew I enjoyed using these habits of mind, I didn't know quite how to teach them.

I can sadly say that despite the importance of teaching the technological design process, almost fifteen years after its emergence in the National Science Education Standards, it remains the stepchild of the science curriculum. In our attempt to "cover" all the content, technological design is often eliminated from the curriculum or merely surfaces at Family Science Nights or at events like Pinewood Derby in scouting.

Sometimes, when I am attempting to explain technological design I refer to a memorable scene in the movie Apollo 13. The movie portrays the 1970 mission to the moon that Captain James Lovell was forced to abort after an oxygen tank explosion damaged electrical systems. The crew had to build a system with the limited supplies on board to remove carbon dioxide. If you saw the movie, I'm sure you recall the scene where NASA engineers on the ground were presented with the challenge of coming up with a solution that the endangered astronauts could implement with limited resources and time. Essentially they had to put a square peg in a round hole with only the materials aboard Apollo 13.

My own story of learning how to engage students in technological design, while not nearly as perilous as the Apollo 13 problem, serves as a starting point for both teachers and parents. When the technological design standard emerged I was determined to learn as much about the process as I could, so I could support teachers in implementing technological design in their classrooms. The first technological design books to hit the market were poor representations of the process. Technological design was presented as a series of "cookbook" recipes with exact steps to follow. I selected one recipe on building simple rockets that was appropriate for third graders and lined up classrooms where I could work directly with teachers to refine my craft and better understand the process of technological design.

My first demonstration lesson could safely be described as dismal at best. The children were all physically engaged. They followed the recipe and built rockets that were equally as inefficient as the model I had prepared in advance. I had accomplished implementing a lesson that could best be described as "hands-on - minds-off" science. All too often I observe lessons that fit this description. The children are engaged in projects where they are "doing" but not "thinking". I am not faulting teachers here, the majority of resources to teach technological design still present it as a series of steps to follow. The instructional materials do not get at the essence of technological design, which is based upon the need to solve a problem or innovate. The NASA engineers, solving the Apollo 13 dilemma did not have a recipe to follow. They had a problem and a limited supply of materials to fix it. As I reflected on my own demonstration lesson on rocket building I knew we could do better than my initial attempt and so I found myself engaged in authentic inquiry focused on improving my own teaching process.

I asked a brand new teacher, whose name was Tom, if he would mind having me work with his students for several days. (Tom told me I could come and teach science for the whole year!) The first day I started in Tom's classroom I abandoned the neat little recipe I had followed in the first demonstration lesson. Instead I sat down with the children and showed them my rocket. I told them I had design issues and wondered if they might be able to offer advice. A deluge of advice followed. "The sides of your rocket are rubbing too much." Ah! Friction! "You aren't pulling the rubber band down far enough." Ah! Stored energy! The comments kept coming. I quickly learned that third graders are in no means bashful when asked to critique an adult's work.

Next, I told the students that they had a week to collect "rocket junk". I told them when I came back we would try to improve on my initial design. I could barely wait for the week to pass. Apparently the children felt the same way as evidenced in the massive pile of rocket junk anxiously waiting for the tinkering of third graders.

Of course we set up some basic management guidelines.

- Develop a list of what you would like to use from the available materials
- Draw an initial design
- Get your materials list approved
- Ask for assistance when you need to use a sharp tool
- Line up at the launch pad when you are ready to test a design
- After testing think about what you want to improve or try next
- Record your thinking
- Continue the process

Tom and I were amazed as we witnessed the engagement and reflective processing of our third graders. By the end of our second day we had rockets hitting the classroom ceiling. On my next visit we moved the whole rocket operation to the gym.

We did not assess the children on their design. Instead, acting as an engineering team, we reflected as a group on design components that were most efficient. These components were listed on charts and illustrated by the children. Our assessment came in the form of an Expository Prompt. "If you could design a rocket again, how would you do it? Support your ideas with evidence! It is important to note that the writing component followed conversation, reflection, and drawing. All too often we ask children to write before they have had sufficient time for processing. Drawing and conversing are ways of thinking. I find that writing improves if we take time for other means of thinking first.

Tom called me the next night. He shared that the expository writing was the best he had ever seen his children do. We suspect it was the active inquiry and experience that led to authentic writing. And so, as a teacher, I had begun to understand what technological design was all about. I invite you to try out the process whether you are a teacher or parent and encourage you to embrace the "messiness" of implementing the process. Have fun!

Technological Design – Continuum Grades K-12

K-2	3-5	6-8	9-12
I made a design plan.	I created and drew a design plan (prototype)	I identified a design problem.	I identified a design problem.
I built my design.	I used science concepts and experiences to create my design.	I determined the criteria for measuring the success of my solution/prototype.	I determined the criteria for measuring the success of my solution/prototype.
I tested my design.	I built my prototype.	I compared alternative solutions by considering the available resources, cost effectiveness, etc.,	I compared alternative solutions by considering the available resources, cost effectiveness, etc.,
I learned from my results.	I tested my prototype.	I designed a prototype based upon my understanding of scientific concepts.	I designed a prototype based upon my understanding of scientific concepts.
I improved my design or decided to start with a new design based upon my test results.	I analyzed my test results.	I built a prototype.	I built a prototype.
	I made improvements or decided to start with a new design based upon my test analysis.	I tested my prototype by collecting the appropriate data.	I tested my prototype by collecting the appropriate data.
		I analyzed my data and evaluate the results.	I analyzed my data and evaluate the results.
		I identified improvements or a new design plan based upon test results.	I identified improvements or a new design plan based upon test results.
		(Please note: The logical process and reasoning involved in technological design is timeless. The challenge and rigor is generated from the design problem and the relationship of the design solution to concepts learned.)	*(Please note: The logistical process and reasoning involved in technological design is timeless. The challenge and rigor is generated from the design problem and the relationship of the design solution to concepts learned.)*

Examples of Technological Design Projects

K-2	3-5	6-8	9-12
Clay boats	Gravity-powered cars	Earthquake resistant structure	Bridge made of paper to support as much weight as possible.
Popsicle savers	Apply knowledge of parallel and series circuits to wiring a box with four compartments (rooms).	Mouse-trap cars	Bridge made of paper to span the greatest distance.
Musical instruments *Science and Technology Curriculum for Children*	Build a home site that can sustain the 100 year flood (Simulation–Stream Tables) *Science and Technology Curriculum for Children*	Rube Goldberg Inventions	Tower made of an unlimited amount of paper to achieve the greatest height.
	Rubber-band powered cars *Science and Technology Curriculum for Children*	Insect monitoring devices	Tower made of a single piece of paper to achieve the greatest height.
	Longest Spinning Top	Macro-invertebrate monitoring devices	Slowest falling parachute made of paper ad string.
	Tallest Newspaper Tower GEMS: *Great Explorations in Math & Science*		Catapult made with wood and rubber bands to shoot a projectile as far as possible.
			Roller coaster made of manila folders and tape to run as long as possible.
			Mobile made of hangers and manila folders to balance as many levels as possible.
			The 9-12 examples were supplied by Aaron Podolner (Golden Apple Award Winning Teacher 2004) Oak Park River Forest High School

FOR TEACHERS

I encourage you to implement at least one technological design project each school year for starters. It is best if you can connect it to your science curriculum as is modeled in the STC (Science and Technology for Children) Program developed at the National Science Resources Center. For example, after children learn about conductors, insulators, parallel and series circuits they are challenged to wire a box with four rooms. You can make this even more creative and authentic by presenting a problem. For example, how could we design a silent alarm for our classroom door so we know when people are coming and going? The idea is to create problems where students can apply the concepts they have learned in science to design technology. **Applying knowledge to new situations requires application level thinking. It requires the transfer of concepts from a learned situation to a problem application. Design technology is a perfect forum for transfer of concepts to novel situations.**

BUT I DON'T HAVE TIME!

Teachers often tell me they do not have time for application level tasks like technological design because the curriculum is so overloaded. My only advice here is this, less is more. I encourage you to start a science curriculum team with the help of your administrators. Work towards creating a scope and sequence with less units covered in greater depth. Teachers find that by limiting the number of units they are required to cover actually gives them the time to investigate in greater depth with greater opportunity for transfer of knowledge to novel situations.

INTEGRATE!

Design technology is a perfect place to integrate literacy with science. Use design technology as a forum for writing across the content area. After children wire a box with four rooms or create a silent alarm have them reflect on the designs that were most effective. Process the information on shared charts. Then challenge the students to do expository writing. Provide a writing prompt. For example, after evaluating classroom alarm designs challenge children to respond to a prompt: *If you could build a silent alarm again how would you do it? Why? Support your ideas with evidence.*

In Barrington Schools we have incorporated technological design into the curriculum at every grade level kindergarten through fifth grades. Teachers tell me that technological deign is highly engaging and really helps them identify kids who have problem solving and engineering skills and habits of mind that are not easily recognized in traditional "book smart" curriculum. The technological design projects the Barrington teachers use are carefully linked with physical science units. The approach is deliberate. The projects are designed in a way that children are invited to apply the concepts that have constructed in physical science to a design problem. For example, after engaging in many investigations of sinking and floating kindergarten children are encouraged to think of a way to get a ball of clay to float. This ultimately leads to something that resembles a boat. Next children are encouraged to extend the carrying capacity of their designs. The process requires shaping, testing, evaluating and modifying designs. It is important to note that the clay boat project was added to the curriculum so that students could apply their knowledge of sinking and floating in a new context, that being boat design.

First graders, in Barrington, are challenged to build popsicle savers. The popsicle saver project culminates a physical science unit where students explore matter through the lens of changes. After a great deal of hands-on experience where children investigate the best place in the classroom to melt an ice cube and various other changes including dissolving and the production of gas, children are challenged to extend their thinking by conducting fair tests using different insulators. Next they analyze insulator data and decide upon a design solution to save a popsicle. Children are given a box and the insulators of their choice to build their popsicle savers. The process continues throughout the grades with increasingly complex problems. Second graders apply their understanding of pitch and volume to musical instrument design. Third graders build gravity-powered cars after investigating force and motion. Fourth graders wire boxes applying their understanding of conductors and insulators, switches and parallel and series circuits. Fifth graders build rubber-band powered cars applying increasing complex concepts of force and motion and energy transfer. The important thing to remember is that the design problems follow the physical science units and afford students with the opportunity to apply learning in new contexts. By now you have observed that as children move through the grades the design problems increase in complexity as the concepts increase in complexity.

When I share the problems elementary children can solve, middle school and high school teachers are often surprised. They often share that these problems are not unlike the ones they have students solve. This brings up the underlying theme of my intent in writing this book. All too often we do not raise the challenge level as children move through the grades. I am not saying that problems cannot be revisited as long as the conceptual understanding is increased. First graders are learning to conduct simple fair tests and apply their test results to design. Middle school students could revisit a project such as this as long as the bar is raised on the thinking. If middle school students are building popsicle savers, then they should be applying their understanding of energy transfer.

USE THE TECHNOLOGICAL DESIGN CONTINUUM TO FOCUS FORMATIVE ASSESSMENT

As mentioned in the "systems thinking" chapter teachers can use the continuums to provide formative feedback as children work through a technological design project. Please see the third through fifth grade section of the Technological Design Continuum played out below.

Continuum Descriptor	Point for Formative Assessment
I created and drew a design plan (prototype)	Does the student design plan make sense?
I used science concepts and experiences to create my design.	Has the student applied science concepts in their design? For example: If they are wiring a four room box have they considered parallel circuits in their design?
I built my prototype.	
I tested my prototype.	Does the student successfully follow their design plan in building a prototype?
I analyzed my test results.	Does the student collect and record data on their design?
I made improvements or decided to start with a new design based upon my test analysis.	Is the student receptive to using their analysis to improve their design?

FOR PARENTS

The technological design process can be part of everyday thinking at home, it works hand in hand with systems thinking. From the early years children can be presented with problems where they have the opportunity to propose solutions, test them out, evaluate the results and modify their plans. It can begin with something as simple as building a tower of blocks. Your job is to simply ask the questions that invite the inquiry. A few examples from block building follow: "How could you make your tower higher? Can you think of a way to connect two towers? How could you balance your tower on a smaller foundation? This kind of questioning can be applied in multiple play situations from the sandbox to the swimming pool.

It is critical that we make the distinction between video games and the kind of technological design problems we are talking about. While video games have their merits they do not require the reflection and application of experience to new design. Technological design problems require perseverance, logic, and innovation. Our children are in dire need of opportunities to build both perseverance and logic. There is no doubt that their world will require innovation.

EMBRACE PROBLEMS

All too often when we are faced with a design problem at home we model our frustration rather than our own interest in finding a solution. In some respects this sends a very clear silent message to children. Problems are to be avoided at all costs! It is absolutely essential that children see us embracing problems. Just as in systems thinking we can model our own problem solving. For example, we seem to have a preponderance of appliances, utensils, etc., that are not made quite as well as they were in the past. When something breaks take a look at it with your child. Invite your child to speculate on how the design could be improved to last longer.

MODEL YOUR OWN CURIOSITY

In our fast paced world, our children rarely get to see us model our own curiosity about how the world works. Take time to engage in conversations with your child about how everyday things work. You can explore the workings of everyday things with your child by visiting the following website http://www.howstuffworks.com/.

HOW CAN PARENTS USE THE CONTINUUM

As with systems thinking, use a grade level band of the continuum to ask questions as your child investigates the solution to a problem. This can be as simple as a gravity-powered car made from everyday recyclables to a car propelled by the air from a balloon. There are a wealth of design technology project ideas on the internet, please see Appendix C for a listing.

Continuum Descriptors	Parent Questions
I created and drew a design plan (prototype)	Did you draw a plan of what you want to try? What materials do you need?
I used science concepts and experiences to create my design.	How did you come up with your plan? What experiences or ideas did you use to come up with your plan?
I built my prototype.	Did you use your design plan to build? Did you change anything?
I tested my prototype.	How are you going to test out your design?
I analyzed my test results.	How did it work? Is there anything you want to change?
I made improvements or decided to start with a new design based upon my test analysis.	How will you improve your design? Do you want to start all over again? What did you learn from your first design?

FAMILY NIGHTS

Help to organize a Family Night at your child's school where family teams can work on design problems together. A few examples follow:

- Build the tallest tower using only newspaper, straws, and masking tape.
- Build a structure out of spaghetti and tape that can hold the most books.
- Use clay, a paper cup, sticky dots, three equal pieces of string and a helium balloon to create the balloon that rises the slowest.

In summary, the thinking and processes involved in solving technological design help children gain a nascent understanding of the nature of engineering. "Engineers solve problems by applying scientific principles to practical ends. They design instruments, machines, structures and systems to accomplish specific ends, and must do so by taking into account limitations imposed by time, money, law, morality, insufficient information and more. In short, engineering has largely to do with the design of technological systems." (AAAS, 1993, p.48) There is no doubt that we need to develop the kind of thinkers and problem-solvers that will embrace the technological challenges we face in the 21st Century. So next time you see a tinkerer in your midst encourage their passion and don't forget to supply the building materials!

Chapter Three: Graphing Data

A graph can be a powerful tool. When we graph data we can discover patterns and relationships. Graphs help us visualize complexities and find the patterns and relationships within those complexities. Patterns play an integral role in mathematics at all grade levels. In *Science For All Americans* (1989) mathematics is described as the science of patterns and relationships. Hyde (2006) believes that when we present mathematics as the science of patterns we have the opportunity "to bring coherence to a bouillabaisse of disconnected ideas" (p.115). He discusses the importance of facilitating learning where students have the opportunity to infer patterns and use those patterns to make predictions. By providing a visual display of data, graphs allow us to see aspects of data, such as patterns, and analyze those patterns more quickly, and in some cases more deeply. This of course beautifully integrates with science. In science, opportunities to facilitate experiences where children have the opportunity to infer a pattern and then strategically use the pattern to make a prediction naturally abound. For example, when primary children graph the data they are collecting throughout the day about shadow length, they can infer that there is a pattern. They love to use their newly discovered pattern to make predictions about how the shadow will progress throughout the remainder of the day or next day. They just need someone who challenges them to make that kind of prediction. Such experiences build children's confidence in using patterns to make predictions.

Our children are growing up in a world that is over flowing with data and information that is readily accessible. Students cannot be interested in the notion of embracing the charge to graph data unless they are connected with the data and understand the relevance of that which they are collecting, graphing and analyzing. All too often I observe lessons where children are required to graph and analyze isolated data on prescriptive forms without any connection to the data or any understanding of its relevance.

Our first challenge as adults is to support our children in becoming critical consumers of the data that is readily available to them. This can be accomplished relatively easily with a series of questions.

- Where did you get the data?
- Can you verify the source?
- Is there other data similar to this?
- What tools did the data collectors use?

- Have others provided conflicting data? If so how does the information differ? Why do you think the results are different?

A series of questions like the ones above allow us to engage in interesting and important conversations with children. These conversations lead to further discussion. For example, we might engage in a discussion about how the data was collected? We can challenge children to think about the concept of collecting additional information and support them in doing so. We can begin this line of thinking at very concrete levels starting at the early grades. For example, if first or second graders are looking at weather data for a given area we can challenge them to compare the data we collect at school or home with the data presented in the local newspaper or on a local weather site. This leads to higher order thinking and promotes a climate of inquiry as students discuss the tools used, the location of the instruments and the accuracy of the data collectors. Once the comparative data is graphed we can engage students in interesting conversations about patterns, relationships and even discrepancies.

At the high school level students often get involved in watershed investigations. Students are challenged to determine stream health by graphing the number of indicator species of macro-invertebrate that are found at a stream riffle. Needless to say the inquiry becomes far more interesting when students overlay other data such as dissolved oxygen level, phosphate levels, temperature and pH. The point being patterns and relationships become readily accessible as complex systems are graphed in relationship to one another.

It is critical that the complexity of the data collection task and the synthesis of data increase as students move from kindergarten to high school. The graphing continuum will help you focus on the critical components within a graph. The reflective language in the continuum encourages children to think of a graph as telling a **story**. Once students realize that graphs tell us stories they begin to construct an understanding of the importance of using graphs, tables and charts. For example, one could graph the sea level depth that krill can be found at varying times of the day. The graph really does not tell an interesting story until we graph whale data on top of that. All of a sudden an interesting story emerges from the data. As krill move up whales move down, better watch out krill!!!!

It is critical that we model our thinking and make transparent to our young data collectors why we use tables and graphs to represent our data. Current technology allows us to graph information almost instantaneously. Sometimes it is easy to make assumptions that children understand the thinking behind the graph that is produced so quickly. The **reality** is this, given the plethora of programs and websites to help us manage our data, we simply plug the data into a computer program and let the program synthesize the information for us. Often we don't even need to consider what to put on a vertical or horizontal axis because the program figures that out for us. **Something is lost in this process!** We have beautiful representations and wonderful graphic displays with little understanding of the relationships displayed. Upon asking probing questions about the computer-generated graph we often discover that students don't fully understand the implications of the data because they don't understand **the thinking behind it**. In *Visual Approaches to Algebra* (1998) students are challenged to figure out which line graph represents a given situation like the rate of cooling of a hot piece of aluminum foil to room temperature. This requires far deeper conceptual understanding. Instead of simply graphing the data and describing the relationship students are challenged to evaluate which graph is the best representation. In this scenario we can assess conceptual understanding of rate of cooling. This is the kind of thinking that we need to work in data rich environments.

As readers who have followed the logic of systems thinking and technological design, you understand that the data we collect helps us look for patterns and relationships and helps us revise our thinking and make further modifications. It gives us something to think about and provides us with the information we need to move forward. Anyone who deals with a great deal of data can verify the importance of representing data graphically. The skilled synthesizer of data looks at the data set and understands which type of graph to use to represent the data most effectively. The skilled synthesizer goes on to look for the patterns and relationships made visible by graphing and then compares that to other patterns and data collected in the past. This is a far cry from what we sometimes challenge children to do in school.

Children have lost the skill set and thinking process described above because we have not challenged them to do so. Since the onset of graphic organizers we have become accustomed to presenting children with tables that are neatly labeled. All the children have to do is fill in the data. We give them graphs with both axes labeled, we tell them that they should use a line graph or a pie chart, we essentially lead them to a schooling of boredom because they don't have to think about which type of graph to produce to display data in a sensical manner. In our well-meaning attempts to support student learning we have merely enabled students to progress on a superficial level without genuine understanding.

The graphing continuum provides a means for teachers and parents to look at the skills and types of graphs students are capable of taking on conceptually as they move through their schooling. All too often, at the beginning of the school year, high school freshmen are merely directed to complete the same seed germination experiments that primary students enjoy. The freshmen are only challenged to collect the same data and represent it graphically. Upon probing, one uncovers the rationale for this alarming lack of increase in complexity. The well-intentioned high school teacher is either unaware of the work done in the elementary grades because they have not seen progressions or rationalizes that they are using the same experiment to teach basic graphing skills. We cannot possibly compete in a global economy unless we raise the bar. Our imperative as adults is to increase the complexity of both data collection and data representation for our children. We must embrace the challenge of looking critically at the complexity of tasks that we give children to do. The continuum below will serve as a starting point. The continuum was based upon a myriad of resources including National Council for Teachers of Mathematics (NCTM) standards to recommendations from the National Science Teacher Association and the National Science Education Standards. It is a blend of recommendations and will benefit over time from the input of others who use it.

Graphing Scientific Data – Continuum Grades K-12

K-2	3-5	6-8	9-12
I put a title on my graph.	I gave my graph a title that communicates what the data shows.	I chose the appropriate type of graph to represent my data.	I chose the appropriate type of graph to display the data.
I labeled the parts of my graph.	I labeled all the parts of my graph.	The title of my graph clearly relates to the information displayed on the graph.	The title of my graph clearly identifies the data displayed on the graph.
I put the data in the right place on the graph.	I made sure I included scientific units of measure.	I used a straight edge or compass to create a neat graph.	I analyzed the range of data to choose the appropriate intervals and sequencing of numbers for the x and y-axes.
I made sure the data on my graph matched our classroom tally.	I put the data in the correct place on my graph.	I chose appropriate intervals to number the x and y-axes. I spaced my numbers evenly on the graph.	The physical intervals on my graph are scaled appropriately and spaced evenly.
I read my graph.	I made sure that the data on my graph matched my data table.	I labeled all parts of my graph correctly: Units of measurement, x and y-axes, columns, rows.	I labeled all parts of my graph correctly: Units of measurement, x and y-axes, columns, rows.
I told what my graph was about.	I wrote a story/description that provides details about my graph.	I placed the manipulated/independent variable on the x-axis and the responding/dependent variable on the y-axis.	I plotted my data set accurately.
I wrote a caption under my graph that tells the story of my graph.		I accurately plotted the data set.	I choose the appropriate line or curve to fit the data.
		I determined trends/patterns in the data.	I explained the meaning of the graph, including patterns and relationships.
		I identified outliers in my data.	I labeled the independent variable on the x-axis and the dependent variable on the y-axis.
		I provided a detailed description and explanation of the data including inferred trends and patterns.	I calculated and interpreted the slope, y and x intercepts.
Data Representations: • Pictures • Tally • Table • Bar graph	Data Representations: • Picture • Tally • Table • Bar graph • Line plot graph • Stem-and-leaf graph	Data Representations: • Table • Bar graph • Histogram • Line plot graph • Stem-and-leaf graph • Circle graph • Frequency distribution	Data Representations: • Table • Bar graph • Histogram • Line plot graph • Stem-and-leaf graph • Circle graph • Frequency distribution • Scatter plots

FOR TEACHERS

MAKE THE DATA INTERESTING

Often times it is easy to stop short of making our data collection interesting enough to engage in conversation above the recall level. For example, we might have students collect data on the number of insects they count on a series of days. We then require that students graph the data and then ask a series of questions about it. Our line of questioning goes something like this. Which day did you find the most insects? Which day did you find the least insects? On what two days was your data most similar? These questions require nothing more than concrete graph interpretation and lack the complexity to engage in thoughtful conversation supported by evidence. Instead of simply collecting data on the number of insects, why not also collect data on weather conditions? When the additional information is graphed we now have a means of using graphs to engage in interesting conversation and inferential thinking. Now we can ask if the students notice any patterns or relationships. For example, students might discover that the warmer the day, the greater the number of insects that they will find. We can then ask further questions to promote a climate of inquiry. I wonder if the time of day impacts the number of insects we would find? How could we find out? Where do you think the insects go on cold days? Where do they go in winter? By broadening the scope of our data collection we will increase the complexity of our own questioning and in turn raise the bar on the thinking our children do.

USE GRAPHS TO INTEGRATE READING & WRITING

As mentioned earlier, graphs are visual representations of data. Each graph tells a story. We can easily integrate reading and writing in the content area with graph interpretation. Synthesis, an important metacognitive reading strategy, can go hand in hand with graph interpretation. By challenging children to place a caption under their graph we are engaging them in synthesizing information. In *Mosaic of Thought,* Keene and Zimmerman point out that proficient readers tend to move beyond the literal meaning of text when they synthesize by employing inferential thinking. As students interpret graphs and synthesize the information they begin to make inferences supported by the concrete data that serves as their evidence. For example, we infer that there is a relationship between temperature and the number of insects found, we use this inference to make further predictions. You can also challenge students to write the accompanying informational text that might appear in a science book next to their graph. Teachers report that expository writing improves when students write to inform based upon their own constructed experiences. You might challenge students to write a "how to" paper about how to collect or how to graph data.

GRAPHS IN INFORMATIONAL TEXT

Informational text is filled with tables, graphs and charts. Model the importance of carefully reading these prior to reading the text. Be certain to point out that these visual representations help us comprehend the text.

DIFFERENTIATE INSTRUCTION WITH GRAPHS

Keep a file of graphs, tables and charts that you find in newspapers, magazines and informational text. Before long you will have a wonderful collection of differentiated instructional materials. When you challenge children to write captions and stories to go along with the graphs you will have a differentiated collection at hand.

USE THE GRAPHING CONTINUUM TO FOCUS FORMATIVE ASSESSMENT

Lets take a middle school example to demonstrate how the continuum can be used to assess formatively. As students work on construction of graphs you can assess many components of their scientific thinking. Take a look at how this might play out on the chart below.

Continuum Descriptor	Possible Student Error	Point for Formative Assessment
I chose the appropriate type of graph to represent my data.	Student selects the wrong type of graph to represent data.	Did the student select a line graph when they are comparing two variables?
I placed the manipulated/ independent variable on the x-axis and the responding/ dependent variable on the y-axis.	Student places the dependent variable on the y-axis.	Does the student confuse the independent and dependent variable?
I determined trends/patterns in the data. I identified outliers in my data.	Student does not pay attention to an outlier in their data. When asked about an outlier they cannot think of possible reasons, for example a measurement error.	Does the student offer possible reasons for having an outlier in their data?

All too often students spend long periods of time constructing graphs and charts without formative feedback. It is critical that students receive formative feedback as they work on graphing. There is no sense spending a great deal of time constructing the wrong type of graph. If a student selects the wrong type of graph to represent data then refer them to the continuum. Ask them to think about the kind of data they are representing. We make it way too easy for students when we tell them which type of graph to use. Challenge them to look at their own data table and go back and find another graph in their book with a similar data set. Challenge them to explain the rationale for choosing the kind of graph they have.

MODEL A PROCESS OF INQUIRY THROUGH QUESTIONING

Often we tell children to produce a graph or table without conveying how graphs and tables help us make sense of the data we have collected. Children are not really certain why tables and graphs are important, they simply know they need to produce them. We tell the children what kind of table to create and provide the labels, we tell them what kind of graph to use, often we provide the template with the intervals all neatly arranged in advance. In so doing we deprive students of the opportunity to think things through. Model your thinking through questioning. It is well worth the time and effort involved. A vignette follows:

- We have been collecting data on temperature for the past two weeks. I am wondering how our data is alike or different from the data in the newspaper.
- I'm wondering how we could compare our data?
- In the past I have used tables and graphs to compare data.
- Do you think it would be a good idea to set up a table?
- What should our table categories be?
- What labels should we use? What should we do with the data on our table?
- I'm wondering if a graph will help us compare our data with the data in the newspaper?
- How can we construct a graph together?
- Do you see similarities and differences between our data and the data in the newspaper?
- Why might our data vary?
- Can you think of a few reasons?
- Can you find any patterns?

It is important to note that when we find discrepancies in data that we should embrace them. The discrepancies are of utmost interest because they hold the potential to advance our problem solving skills. As we consider alternatives students often discover that their errors lead to deeper understandings and improved methodologies.

FOR PARENTS

Graphs, charts and tables are simply part of living in the 21st Century. As adults we often rely on them when we skim through a newspaper or business magazine. Take the time to talk to your children about ways in which you rely on graphs and tables when you are reading. Share ways you use graphs at work. Cut graphs out of the paper and remove the title and caption. Challenge your child to come up with a great title for the graph. Challenge your child to think of an interesting caption for the graph.

Help Build Connections

As you look at graphs with your child build connections by challenging them to think of another time when they have seen a similar pattern or representation. Challenge them to think of something else that could be represented by a similar graph. For example, there are many graphs that represent direct relationships where as one thing increases the other increases or where as one thing decreases the other decreases. Some examples follow:

- As the temperature increases the number of cricket chirps increases.
- Thinking in a time frame from noon to sunset, as the hour increases the length of a shadow increases
- When we increase the number of workers we increase the production of goods
- Accidents increase when texting increases

- Accidents decrease when texting decreases
- If you decrease your calorie intake you will decrease your weight

It is fun to use challenges like the one above as topics for family dinner conversation or on rides in the car. Taking the time to think about intelligent conversations to have at the dinner table increases the level of conversation and in turn the thinking and communication skills your child will develop.

Question, question, question!

If your child produces a graph on the computer, then ask questions to be certain they understand that which was produced. Challenge them to explain the relationships or patterns they see. Ask them to make predictions based upon the pattern that emerged and by all means resist the temptation to create the graph for them. It is not your homework!

Chapter Four: Scientific Investigation

The terms and circumstances of human existence can be expected to change radically during the next human life span. Science, mathematics, and technology will be at the center of the change -causing it, shaping it, responding to it. Therefore, they will be essential to the education of today's children for tomorrow's world.
Benchmarks for Science Literacy, 1993

When challenged to recall their own science education most adults make reference to what we traditionally called the scientific method. They reminisce about a rigid process and often share some of the steps they followed like, state a hypothesis, list your materials, follow a series of steps and so on. Science, for many of us, was presented as copious volumes of facts to be recalled and methodical steps to be followed. The traditional teaching of science left many of us with misconceptions about what science is. If we were to ask a scientist, we would discover that the scientific community describes science as a highly creative endeavor, an intricate blending of imagination and logic. (Science For All Americans, 1989) Science is about rich conversation that engages scientists in thoughtful discussion about claims that are supported by evidence. The evidence emerges, not from minds that see the ordinary but from those that choose to be open to the extraordinary. Science, for children, should be engaging and should be driven by questions and opportunities to investigate the wonders and complexities in our world. It should emphasize the role of evidence in making claims and help children understand that the possibilities are only limited when we fail to ask the important questions that may lead to new evidence. Science is a dynamic enterprise that requires integrity, diligence, fairness, curiosity, skepticism, comfort with ambiguity, imagination, and the willingness to consider new evidence.

It might come as a surprise to many of us that when we solely think of science as direct experimentation, our view of science is poorly illuminated. From simulations, to field studies, to theory testing, to model building scientists employ a vast array of practices and use a myriad of tools. They accept ambiguity fully understanding that new evidence may serve as cause for reconsidering current claims. For example, in 1909, when Einstein presented his first lecture in Salzburg he was the first to clearly introduce the concept that light had dual properties. Scientists before him had explained light as either a wave or a particle. Einstein's concept of duality provided new evidence and an opportunity for other scientists to reconsider their own conceptual understanding. I highly recommend Joy Hakim's series *The Story of Science.* Her books engage students in the wonder of science and celebrate the imagination, creativity, and resiliency of scientists. In her third volume *Einstein Adds A New Dimension* readers get a full appreciation of the hard-won knowledge and sense of wonder that prevailed as Einstein added a new dimension to scientific thinking.

Children should have opportunities to read about the dynamic work of scientists in the field. There are many wonderful children's books that introduce children to the intriguing work of scientists. The scientists celebrated in these books work in many different places and settings using different methodologies and protocols. For example, *Extreme Scientists: Exploring Nature's Mysteries From Perilous Places,* authored by Donna Jackson introduces children to adventurous scientists from hurricane hunters to microbiologists who work in caves deep below the earth' surface. I have included a list of some of my favorite books about scientists for both classroom and home libraries as Appendix D.

It is important to recognize that the Scientific Investigation continuum only presents one set of practices that scientists use. The reflection points are worth thinking about while fully recognizing that there is not one locked step method involved in doing science. Nevertheless it is important for students to understand that scientists do engage in sharing their claims with others and must supply the supporting evidence. At times, given the nature of the question, it is important to set up control groups and make evidenced based observations. At times it is important to use a series of logical steps that others can follow. All too often, based upon our own experiences in school we present children with lock steps, we challenge them to formulate predictions and/or hypothesis statements without prior knowledge or experience. Imagine the thinking that we can stimulate if we take the time to uncover prior knowledge and use it to support children in formulating questions and hypotheses based upon their prior knowledge and experiences. The continuum will provide guidance for both you and your children in reflecting on the process of scientific investigation. Please remember that it is a dynamic process that requires cycle after cycle of investigating and questioning. Each investigation we do should serve as the catalyst for new questions and new discoveries.

MISCONCEPTIONS

Let's begin with one of the salient principles of learning, highlighted in *How Students Learn* (2005). "Students come to our classroom with preconceptions about how the world works. If their initial understanding is not engaged, they may fail to grasp the new concept and information, or they may learn for the purposes of the test but revert to their preconceptions outside the classroom."(p.1) Our challenge, as teachers and parents, is clearly to uncover the misunderstandings children bring to the table. Many of us have laughed at the responses Jay Leno gets when he asks simple questions of average Americans on the street. While on one level I suppose we can think of the responses as funny, on another level they make me fully aware of the lack of progress we have made in science education. For example, as well document by Annenberg Foundation in their documentary *Private Universe,* many of us reach adulthood with scientific misconceptions. As highlighted in the documentary through interviews, when challenged to explain seasonal change many adults equate seasonal change to distance from the sun rather than the tilt of the earth on its axis.

Both the Barrington and River Forest teachers find it helpful to use an ICE (Ideas, Claims & Evidence) chart as they uncover and work through student misconceptions. The chart is used throughout a unit of study and receives on-going attention through conversation in which students compare the evidence they are collecting with their initial ideas. A sample of the ICE Chart appears below highlighting a simple misconception children hold regarding magnets.

Ideas My beginning ideas… What I think might happen	**Claims** What I found out	**Evidence** Data and/or observations that support my claim
All metals are attracted to magnets.	Not all metals are attracted to magnets.	The paper clip, nail, pin, scissors were attracted to the magnet. The penny, copper wire, and earring were not attracted to the magnet.

The ICE chart works best when we create a classroom environment where children feel comfortable sharing their initial ideas. The initial ideas serve as the beginning point for inquiry. Let's play this out using the sample ICE chart above. We have chosen a simple example to explain the ICE process. Needless to say the process increases in complexity when we get to more complicated misconceptions that require greater investigation and evidence. You might begin an investigation of magnets by asking children where they have seen magnets or you might simply provide an opportunity for some open-ended exploration with magnets and materials. Next you might ask the students to share their initial ideas and findings. Initial ideas are posted on the ICE chart. Typically common misconceptions emerge at this point. As you can see on the ICE chart above in the "Ideas" category, the initial "idea" about magnets reveals the misconception that all metals are attracted to magnets. Next you would challenge children to think of a way to figure out if "all metals" are attracted to magnets and provide the materials for exploration. Of course you would encourage the children to sort materials into two piles and would probably suggest that they set up a table to record their findings. On the classroom ICE chart student findings are placed under the "Evidence" category. Now the important metacognitive processing begins. Challenge children to come up with a "Claim" based upon the "Evidence" they have gathered. After the "Claim" is formulated encourage the children to compare their "claim" with their initial idea. Challenge them to synthesize by asking: "What could we tell someone who believes that all metals are attracted to magnets? What "evidence" would you give them? What could you have them do? The important thing is to compare the "claim" with the "idea" and insist on "evidence" when children supply their explanation.

The ICE Chart is a blend of the work on conceptual change theory, initiated at Cornell University (Posner, Strike, Hewson, & Gertzog, 1982), with more recent works that highlight claims and evidence set forth in the National Science Education Standards (1996). It builds metacognitive thinking into Ogle's (1986) KWL reading strategy. ICE supports children

33

in developing analysis, synthesis and evaluation strategies as they analyze data, synthesize evidence into claims and evaluate their claims in relationship to their initial ideas.

In summary, when students use ICE to organize thinking they use their initial "ideas" to formulate questions for research and/or to design investigations. "Evidence" is gathered through careful observation and research including quantitative and qualitative data. Learners gain experience organizing and representing their data in meaningful ways so it is readily available for analysis. "Claims" are based upon the evidence. Learners are encouraged to write and draw explaining how their ideas have changed. The process is deliberate and on going. Teachers tell me it really encourages children to slow down and think about their thinking. They also tell me that ICE provides a wonderful gateway to persuasive writing. We will revisit ICE in the last chapter on the integration of reading and writing with science.

Sometimes teachers ask why the "Claim" section appears in the middle of the chart. The rationale is based upon the research in *How People Learn* (2000). If children have a misconception and we teach them something new, they will learn the new information for the test and frequently revert to the misconception afterwards. By placing initial ideas next to claims children can readily compare the two. Typically students keep a journal with their evidence. The chart is completed as a processing tool to synthesize the hands-on investigations, data collection, research and journaling.

SCIENTIFIC INVESTIGATIONS

Many of the scientific investigations available to teachers and parents still focus on a set of steps and procedures to follow. The key is to take the investigations provided to the highest level possible by modeling our own interest in moving beyond the activity. For example, an initial investigation might challenge students to compare the rate of fluid flow using different liquids. The typical investigation tells the teacher and student exactly how to proceed. Such investigations can serve as the seeds to inquiry. Try to extend the thinking through questions like these: What would happen if we changed the temperature of the liquid? What would happen if we used different surfaces? What would happen if we changed the slope of the surface? By now you have the idea, take any investigation and start modeling your own wondering. Best practices in reading and writing emphasize the importance of modeling our own reflective thinking. Modeling is equally as important in science, social studies and math. Once we model our own interest in generating new questions and investigations, it does not take long for inquisitive children to catch on and come up with additional questions and investigations they would like to try. This leads us to an important point. All too often we race from one investigation to the next instead of slowing down and exploring the concept in depth. Give yourself the luxury of time to go in depth. You will be amazed by the questions and investigations that will emerge from one simple inquiry.

Scientific Investigation – Continuum Grades K-12

K-2	3-5	6-8	9-12
I started my fair test with a question.	I started my investigation with a question I could test.	I started my investigation with a testable question.	I started my investigation with a testable question.
I made a prediction after thinking about what I already know.	I made a prediction/hypothesis based upon things I've read and observed.	I developed a hypothesis that is based upon my research, observations, and previous scientific investigations.	I developed a hypothesis that is based upon my research, observations, and previous scientific investigations.
I wrote down/drew the one thing I would change.	I wrote my prediction/hypothesis using my science language.	I wrote my hypothesis as a testable cause and effect statement.	I wrote my hypothesis as a testable cause and effect statement.
I listed/drew all the things I would keep the same.	"I think _____ will happen because…" or "If _____ then _____ because…"	I designed an experiment that effectively uses controlled, manipulated/independent and responding/dependent variables.	I designed an experiment that effectively uses controlled, independent, and dependent variables.
I decided what I would observe or measure.	I understand that some predictions may no be written as "if" "then" statements.	I listed the materials I need to conduct my investigation.	I listed the materials I need to conduct my investigation.
I made a plan to record what I observed.	I identified the manipulated variable/ the one thing I would change.	I developed logical steps/procedures to follow. My steps are written clearly. Someone else could easily follow my steps/procedures.	My procedures are clearly communicated.
I recorded what I observed using drawings, numbers and words.	I identified the controlled variables/constants/ all the things I would keep the same.	I recorded quantitative and/or qualitative observations/data.	I collected data that uses quantitative and/or qualitative measures.
I looked at what I recorded and talked about it.	I wrote a step by step plan to follow.	My sample size was sufficient.	I displayed my data using appropriate graphic formats.
I made a claim/conclusion.	I tested my plan.	I displayed my data using appropriate charts, tables and graphs.	I manipulated and analyzed my data by using appropriate statistical measurements.
I listed my evidence.	I recorded my data using both observations and measurements.	I used my data to tell a story. I developed claims/conclusions that are supported by evidence/data.	I developed claims/conclusions that were based upon evidence.
I compared my claims and evidence with other scientists.	I analyzed/looked at my data and talked about it with other scientists in my class.		
I asked a new question.			

K-2	3-5	6-8	9-12
	I made a claim based upon my evidence.	I can identify other factors that may be influencing my results. (Potential Error)	I used scientific terminology when reporting claims. My terminology is accurate and appropriately used.
	I supported my claim with evidence.	I recognize and can explain how different explanations can be given for the same results. (Potential Bias)	I determined whether differences in results were trivial or significant. My judgments are supported with evidence.
	I compared my claim with others after reading and discussing my claim with others.	I do additional research to verify my claims.	I reflected on ways in which others might explain the data recognizing that more than one explanation is possible.
	I discussed differences in my findings and the findings of others.	I state the limitations in my preliminary findings.	I proposed and/or conducted additional research to support or negate claims/conclusions drawn.
	I recognize that more than one explanation can be given for the same observations	I identify new questions for investigation based upon my findings and conversations with other scientists.	I identified legitimate sources of uncertainty.
	I reflected on how my ideas changed and developed a new question.		I recognized that further observations were needed to observe patterns.
			I identified new questions for investigation based upon my findings and conversations with other scientists.

FOR TEACHERS

CREATE CLASSROOM CHARTS

As mentioned in chapter one, create a chart of the continuum expectations for your grade level band. This creates a focal point for instruction, clarification of expectation and most importantly student reflection. A sample classroom chart for grades 6–8 appears below.

Scientific Investigation

I started my investigation with a testable question.

I developed a hypothesis that is based upon my research, observations, and previous scientific investigations.

I wrote my hypothesis as a testable cause and effect statement.

I designed an experiment that effectively uses controlled, manipulated/independent and responding/dependent variables.

I listed the materials I need to conduct my investigation.

I developed logical steps/procedures to follow. My steps are written clearly. Someone else could easily follow my steps/procedures.

I recorded quantitative and/or qualitative observations/data.

My sample size was sufficient.

I displayed my data using appropriate charts, tables and graphs.

I used my data to tell a story. I developed claims/conclusions that are supported by evidence/data.

I can identify other factors that may be influencing my results. (Potential Error)

I recognize and can explain how different explanations can be given for the same results. (Potential Bias)

I do additional research to verify my claims.

I state the limitations in my preliminary findings.

I identify new questions for investigation based upon my findings and conversations with other scientists.

HYPOTHESIS DEVELOPMENT

Let's take time to consider a few of the expectations above. The ideas can be readily applied to other grades. Let's start with hypothesis development. Many of us can recall experiences in school when we were asked to develop a hypothesis statement without prior knowledge. We diligently responded to the directive and often developed a hypothesis that would have been better described as a "guess". When we ask children to develop a hypothesis without prior background or experiences we are perpetuating misconceptions about the way in which scientists go about their work. A well-developed hypothesis is dependent upon a significant body of research, observation, and/or previous scientific investigations. In *Taking Science to School* (2007) readers are challenged to consider the development of questions and hypotheses as part of an "iterative cycle" (p.131) rather than as the first step in the scientific method. Simply stated, hypotheses statements and further investigations should grow out of previous cycles of inquiry. The hypothesis statement a student develops should incorporate previous experience and understandings. Please recognize that the placement of question and hypothesis on the continuum does not infer a step by step methodology. In *Taking Science to School* (2007) we are reminded that observations gathered through experimentation should minimally "serve as evidence that will be related to hypotheses" (p.131).

IDENTIFYING NEW QUESTIONS

Teachers will often ask why the identification of new questions appears again at the end of the continuum. This is purposeful. It represents the continuous, dynamic nature of inquiry. *The asking of new questions is what it is all about.* We want children to leave our classrooms asking more questions, we want their curiosity to continue to grow. Furthermore, the evidence gained in a single investigation should serve as the seed for future question asking and hypothesis formulation. Simply stated, one thing leads to another.

HABITS OF MIND

The last five reflective statements on the 6-8 Classroom Continuum Chart may have the most potential for simulating the authentic work of the scientist. They challenge students to consider potential bias, potential error and to verify their claims with evidence. They encourage students to consider limitations and to ask new questions. These are habits of mind that will serve our children well as they deal with the deluge of information that is literally at their fingertips. Often times, in our well-intentioned desire to cover the curriculum, we fail to spend enough time on these habits of mind. Please consider doing less investigations, but doing the ones you do in enough depth that students have the opportunity to engage in conversations regarding potential bias, potential error and limitations. Be certain to carry out an entire investigative cycle ending with new questions for further investigation. Encourage them to consider the credibility of claims, as you ask them to consider sample size and methodology. Challenge them to consider what additional information they might need to draw a conclusion. Recently, in reviewing sample assessment items from The Programme for International Student Assessment (PISA), it became clear that internationally students are challenged to apply the habits of mind we are discussing here as they respond to test questions.

DIFFERENTIATE INSTRUCTION BASED UPON STUDENT MISCONCEPTIONS

Please try out the ICE Chart to determine initial ideas that your students might have related to science concepts. It takes lots of probing to get at misconceptions. Most current science programs provide common misconception alerts for teachers. Be certain to read them so you can focus the questions you ask to get at common misconceptions. Once you are aware of the misconceptions your students have differentiate to address those misconceptions by supporting students in framing questions and investigations to confront their misconceptions. As students investigate encourage them to compare the evidence they are gathering through their investigation with their initial ideas. Specifically ask them how their new found evidence compares with their initial ideas.

USE ASSESSMENT PROBES

NSTA Press has published a series of formative assessment probes that address the general misconceptions that many students hold. Title: *Uncovering Student Ideas in Science,* lead author Page Keeley. Please see the reference section at the end of the book for the full reference. Each probe is followed by alternative responses. The alternative responses are designed to target common misconceptions. As you work through the unit you can challenge your students to revisit their initial response given the new evidence they have gained through classroom investigations and readings.

Please also refer to the last chapter in *Benchmarks for Science Literacy, The Research Base* (1993). This gem of a chapter discusses many of the common misconceptions we hold regarding science concepts and the nature of science. It is a gift to every teacher and parent interested in thinking about the kinds of questions they might ask to find out what children are thinking. For example, in The Research Base we learn about the classic studies of Freyberg and Osborne (1985) who determined that children do not view trees as plants. Once we know that children may not consider a tree to be a plant, we can ask our own probing questions. What is a tree? How is it like and different from a bean plant? Is a tree a plant?

By reading the research base we can learn about common misconceptions children hold and ask probing questions to get at their thinking.

FOR PARENTS

Perhaps the best way for parents to encourage the kind of thinking that undergirds scientific investigations is by modeling your own skepticism when you read about a study or hear a claim on television regarding the benefits of using a certain product. The modeling might go something like this:

"I just read a really interesting article about the positive benefits of drinking pomegranate juice. I wonder how they determined that it was the juice that showed health benefits? I wonder who conducted the survey? It seems like with all the other factors it would be hard to know if it was really the juice. Where do you think we could find more information? How could we check the reliability of the sources?

The idea here is to model your own thinking process. It is almost impossible to live a single day without hearing about the benefits or risks involved in something we do. Let your children see you think it through before you accept or reject the claim. It is perfectly fine by the way to state that you are not convinced and would like to gather more information.

USE ASSESSMENT PROBES TO START DINNER CONVERSATIONS

As mentioned in the teacher section above, NSTA Press published a series of formative assessment probes that address the general misconceptions that many students hold, as adults we hold some of the same common misconceptions. (By formative assessment probe we mean a question or prompt given to students to determine their prior knowledge and thinking.) The series of books are titled *Uncovering Student Ideas in Science,* lead author Page Keeley. Please see the reference section at the end of the book for the full reference. Each probe is followed by alternative responses. The alternative responses are designed to target common misconceptions. Start a dinner conversation by asking family members to commit to an answer based upon the probe and then engage in discussion. The book provides the "best response" to each probe and provides clear explanations to help even the most science shy reader understand the common misconceptions they might hold. Dinner conversations will be lively and the whole family will have the chance to become more scientifically literate in the process.

SCIENCE FAIR

Perhaps it was just luck of the draw, but in all my years of judging Science Fairs only one project stands out. For me, it was because the initial inquiry came out of an authentic question that generated interest and a wonderful process of inquiry. The young man standing beside his science fair project introduced himself and then immediately launched into an explanation of NFL rules regarding the pressure of game day footballs. He went on to share that kickers used footballs dedicated solely for the kicking game. He shared that he wondered about the ruling and decided that he wanted to figure this all out. He and his grandfather set out to build a wooden leg on a pendulum with a consistent starting point and force to be applied. He shared how he manipulated the pressure of the football and ran a series of trials over several days comparing the distances the football traveled under varying pressure. He shared that one very cold Saturday morning, when he went out to conduct another series of trials, he was amazed at how much the pressure of the ball had changed over night. This of course led him to an exploration of the relationships among volume, temperature and pressure. While the experimental design was rather simplistic the constructed conceptual understanding of physical science concepts was what interested me. In truth, this story is told to get at a simple point. If students do science fair, then they should be genuinely interested in what they are investigating. They should not be developing hypotheses without observation and evidence gained. Their hypothesis should only come after cycles of inquiry. Their questions should stem from observation of the real world and the integration of science concepts in everyday life.

Science will continue to play an increasingly important role in our lives as we move through the 21st Century. Our children will need to be critical consumers of all the information that is so readily accessible. They will need to understand that the information they receive is dependent upon the reliability of the data collected and the methodologies used to collect it. The imperative is to nurture the kind of thinkers who look carefully at the information they gather and its multiple sources before jumping to conclusions. Modeling our own interest in doing so is essential.

Chapter Five: Scientific Drawing

Students need not only to do hands-on science and talk and write
science in words; they also need to draw, tabulate, graph, geometrize,
and algebrize science in all possible combinations.
Lemke, 2004

Many scientists have been described as visual-spatial thinkers. "Michael Faraday's visualization of lines of force surrounding charged objects and magnetic poles is frequently cited as an example of the use of imagery in explanations of phenomena" (Mathewson, 1999, p.37). In *Beautiful Evidence* Tufte discusses the significance of the annotations that accompanied Galileo's scaled images of his observations of the moons of Jupiter. "Because of the detailed annotations, the drawings became credible quantitative evidence about satellite motion, not merely still-land sketches of telescopic views" (Tufte, 2006, p.13). Drawing is a way of thinking, reflecting, and processing information. It supports the development of visual-spatial thinking. Often we fast forward students from a hands-on investigation to a writing exercise in which we ask them to incorporate vocabulary and new concepts that may stand in contrast to their misconceptions. We miss the opportunity to afford students with an opportunity to process their experiences and to talk about them before we engage them in writing. We miss the opportunity to have them label drawings, reinforcing newly introduced vocabulary and, most importantly, we miss the opportunity to provide the quintessential processing time that comes before writing. Drawing and talking about experiences provides cognitive rest and often allows details to emerge that reveal misconceptions. For example, a seemingly adept student might parrot back an explanation of seasonal change only to have their drawing reveal a misconception in which they illustrate a geocentric model in which the sun revolves around the earth. Drawing and talking inform understanding in science and drawings often reveal student understandings in need of further clarification experiences. As students draw and journal we have the perfect opportunity to engage in important conversations about their work.

If you are not convinced that a box of crayons and a drawing surface can lead to interesting conversation just take time to observe family interactions in a restaurant where crayons and a drawing surface are supplied. Before you know it everyone is holding crayons and either adding to the initial drawing or turning it into something quite different. It can be a delightful interchange of ideas and illuminates the timeless significance of family time.

In *Crossing Borders in Literacy and Science Instruction,* Lemke (2004) encourages the literacy experts to learn from the multimodal representations that scientists use to convey meaning. He states: "Text is becoming more and more integrated into multimedia. Purely textual literacy will survive, but it will not continue to hold the dominant place it has in the past" (p.42). Lemke points out that students will need to know how to look critically at animations, schematic drawings and diagrams to mention a few. Supporting students in understanding the nuances of scientific drawing should certainly support the kinds of visual literacy skills they will need to process a bombardment of information and images so readily accessible. There is no doubt that visual images, schematics, and diagrams will take center stage in 21st Century learning. All one has to do is open a scientific journal to fully appreciate the multimodal language that scientists use.

The scientific drawing continuum prompts students to reflect on their own scientific drawings. It encourages them to slow down and add greater observational detail. The kind of detail that leads to building the observational skills needed to uncover the intricacies in systems great and small.

Scientific Drawing – Continuum Grades K-12

K-2	3-5	6-8	9-12
I drew what I saw.	My drawing looks similar to what I observed.	My drawing is realistic. It depicts the object observed.	My drawing is detailed and includes both quantitative and qualitative observations.
I labeled all the parts of the drawing.	I included as many details as possible: color, textures, shapes, measurements, etc.,	I have included many details: size (with metric measurements), colors, textures, shapes, and relationships to surroundings.	My drawing includes only the features I actually observed and not features I inferred.
I gave the drawing a title.	I labeled all the parts of my scientific drawing.	My title is descriptive and accurate.	My title is descriptive and accurate.
I made the drawing clear enough to see all the parts.	My drawing is neat and easy to follow.	My drawing is neat and can be easily read.	All parts of my drawing are clearly and accurately drawn and labeled.
My drawing is neat and easy to follow.	I wrote a title that tells what my scientific drawing shows.	I clearly and accurately labeled all the parts of my drawing.	I drew multiple perspectives to provide the viewer with a complete picture of the object under study.
I used my science words to label my drawing.	I wrote an explanation of what my scientific drawing is trying to show.	I have drawn multiple perspectives to provide the viewer with a complete picture of the object under study.	I included a detailed explanation of what the scientific drawing is intended to show.
	I used my science vocabulary in my explanation.	I included a detailed explanation of what the scientific drawing is intended to show.	I used scientific vocabulary throughout my explanation.
	My scientific drawing is large enough for details to be recognized.	I used scientific vocabulary throughout my explanation.	I used a very precise scale and proportions consistently throughout my drawing.
		My scientific drawing is of an appropriate size and scale for details to be recognized, or magnification is indicated.	I accurately described my metric scale.

Anne Grall Reichel, Ed.D.

FOR TEACHERS

Set The Stage For Observation Through Questioning

Last fall I had the opportunity to observe a group of students on a field trip. They were charged with finding, observing, drawing and taking digital images of insects. As the children charged off the bus the only thing they seemed to engage in was reckless abandonment of all goals at hand. It seemed that they were so accustomed to rushing and instant gratification that they took off down the path at lightning speed, only to return within five minutes to tell their teacher, quite confidently, that there was not a single insect to be observed along the entire path. We need to slow children down before they can observe let alone draw. Before you take children outdoors to observe please have important conversations about the observation task. Your questioning can go something like this: Where do you think you might hide if you were an insect? What color might you be? Do you think you are more likely to find insects closer to the ground or under a leaf? What kind of insects might you find inside of flowers? Do you think different types of insects might be in different places? Setting the stage for observation through questioning is well worth the time and conversation prior to field experiences. You might try sitting students down quietly along a path and ask them to focus on one plant or one small square of ground. This quieting is essential to good scientific observation and to the drawing that subsequently follows.

Scientific Drawing as Formative Assessment

As children work on the details in their scientific drawings we have the opportunity to ask probing questions about their understandings. For example, if students are drawing about their experiences with shadows we should be able to determine if they really have constructed an understanding of where a light source, an opaque object and a shadow would be in relationship to one another. As students draw a picture representing sun, moon and earth we can determine if they really understand what orbits around what. We can challenge them to describe rotation and revolution, we can ask them to show where the sun, moon and earth would be in relationship to each other if we were looking at a full moon. A classic food chain often reveals misconceptions about energy flow. Children will often mistakenly draw an arrow from the organism doing the eating to the organism being eaten. In this scenario their understanding of energy flow is in need of further modification. The most important thing is that we should be attentive and ever present as children are engaged in scientific drawing, because it is a perfect opportunity to have conversations about the science concepts and to challenge students to tell us more about their thinking.

We can have questions in mind based upon our continuum level to support students in reflecting on their progress with scientific drawing. Please see the example below.

Continuum Descriptors	Question
My drawing looks similar to what I observed.	What could you do to make your drawing look like the _____ you observed?
I included as many details as possible: color, textures, shapes, measurements, etc.,	Are you missing any details? Is there a tool you could use to add detail? (i.e. ruler or hand lens)
I labeled all the parts of my scientific drawing.	Did you use all your science words to label your drawing?
My drawing is neat and easy to follow.	Would someone be able to understand your drawing? What would make it more understandable?
I wrote a title that tells what my scientific drawing shows.	Does your title convey your thinking?
I wrote an explanation of what my scientific drawing is trying to show.	Did you write a caption or explanation so others know what you were thinking?
I used my science vocabulary in my explanation.	Did you use your science vocabulary in your explanation?
My scientific drawing is large enough for details to be recognized	Would it help to change the size of your drawing so we can see more details?

Scientific Drawings and Vocabulary Development

A few years back, while working on formative assessments with the Barrington teachers, we discovered an important scaffold that we could provide for students who were finding it difficult to transfer words from the word wall to their drawings. We started placing the vocabulary words we wanted children to use directly under the boxes on journal pages. For our ELL students we gave them envelopes with the vocabulary words in English on one side and in Spanish on the other. We asked them to place the words on their drawings and challenged them to use the words in their explanations. The matching of science vocabulary to constructed experiences and drawings supported all learners in building conceptual vocabulary and in making the transition to writing.

Scientific Drawing as a Scaffold to Writing

As mentioned in the introduction to this chapter, drawing is a way of thinking. As Marzano & Pickering (2005) point out drawing involves non-linguistic rather than linguistic representation. As children draw they naturally converse about their drawings and this in turn helps them make the transition between non-linguistic and linguistic representations. The drawing and conversation provide a scaffold to support students in writing scientifically. The labeling of scientific drawings reinforces vocabulary and seems to support children in constructing meaning. The research supports our observations. Pearson and Cervetti (2005) found that students needed to hear words many times before they can incorporate them in speech and writing. The drawing and labeling process provides additional context prior to writing. Besides the informal discussion serves an important purpose, it allows students to share their representations and "practice" the "talk" that goes with their drawing.

FOR PARENTS

Slowing Down to Discover Something You Have Not Seen Before: The Wonder Of It All

Earlier in the book you read about the importance of reconnecting children with nature. If you only try one suggestion in the book, I really hope it is this one. From symmetry to intricate detail, nature provides a rich forum for observation. Challenge your child to draw from memory one thing in nature that they have often passed by while walking or running. The object could be a tree, a flower, a leaf, or even a blade of grass, just something outdoors. The lack of detail will probably be astounding. Now tell your child they can examine the object of choice for two minutes and challenge them to draw it again. Finally give your child a magnifying lens to look at the object. (Every child should be equipped with their own magnifying lens when they are experiencing nature.) Challenge them to draw the object again adding details discovered with the hand lens. Both you and your child will be astonished at the newly discovered details that will emerge in the drawings. The point here is that we need to slow kids down. The wonder and detail in both the designed and natural world is really quite amazing if we take time to slow down long enough to see it. Observe the side of an old penny with the Lincoln monument, you will be fascinated by the details you never saw. The scientist attends to detail in her observations and drawings. It is in the discovery of the intricate details that we learn to better understand the system as a whole.

Continuums & Questions

Use the continuums to create questions that will support your child in reflecting upon the quality of their drawings. Please take a look at the example below.

Continuum Descriptors	Questions
I drew what I saw.	Take a look at the _____ again. Did you discover anything else? Can you add any details to your drawing?
I labeled all the parts of the drawing.	Do you need help labeling your drawing? Do you know what these parts are called?
I gave the drawing a title.	What do you think a good title would be for your drawing?
I made the drawing clear enough to see all the parts.	You might want to look at the _____ again. Are you missing any parts in your drawing?

Technical Drawings

Our world is filled with technical drawings and schematics. We use them to assemble bicycles, set up computers and lawn furniture. Our list could go on and on. Often Dad gets the job. The rest of the family is cautioned that quiet must prevail while Dad is concentrating. Sometimes construction involves an explicative or two. This sends several very clear messages. It is Dad's job to follow schematics. Technical drawings and schematics are hard. Girls can't do this job and so on. Girls should be challenged to work with schematics and technical drawings. As parents, it is our responsibility to expose them to the intellectual challenge and support them in building their confidence with tasks that require them to use technical drawings. While this next suggestion might seem quite burdensome it is not a bad idea if children see their Mom tackle the task or perhaps Mom and Dad model their thinking as they tackle it together. Our children need to see us embrace these challenges. The use of images, icons, technical drawings and schematics will only increase in our highly visual society. Children need to hear and see us embrace the technical drawings so they don't fear them.

Chapter Six: The Essential Connection: Writing & Reading in Science & Social Studies

Primary-grade students who do more writing, especially **content-related writing**, have higher reading achievement. Although writing connected to reading is the single best indicator of student achievement, such writing is not common.
Regie Routman

The best advice seasoned educators can pass on to the next generations of teachers and parents can be summed up in one simple yet complex word. If you have synthesized Regie Routman's pearls of wisdom above, well supported by research, and taken in the spirit of the preceding chapters you have probably already inferred what that word might be. The word is "integrate". If we do not integrate the content we teach in science and social studies with reading and writing we will inevitably drown in perpetual coverage of disconnected thoughts and ideas and run the risk of completely losing a generation who can access a world of information instantaneously. Integration of curriculum around meaningful content frees us to plan logical curriculum and provides interesting content for our children to read and write about. The content of science and social studies provides the forum for "writing to inform" and "writing to argue," two of the charges in the new Common Core Standards. Information is readily accessible to our children, the imperative is providing the skills for them to process the information intelligently and then communicate it articulately moving from text to writing to visual images effectively. Every adult needs to understand that our day may still be approximately 24 hours but obviously the world our children live in is spinning much faster. Our traditional approaches are simply not enough.

For well over twenty years literacy experts have been talking about the importance of valuing reading and writing as processes that support one another. In 1983, Stotsky pointed out that the students who were most proficient at writing were also proficient at reading. Reading and writing are inextricably connected. Many writers talk about reading a great deal when they need to get back into their writing zones. That said; this research has important implications for both teachers and parents. We need to create text rich environments for children. The

Common Core Standards recommend that by fourth grade children should be spending 50% of their reading time reading informational text. So the next time you order books for your classroom or home library make sure you have a balance of classic literature and informational text.

There is a significant body of research that emphasizes the importance of nonfiction writing. Reeves (2002) showed that when we increase nonfiction writing by a mere 15 minutes per day that it has a significant impact on student achievement. He emphasized the importance of providing specific feedback to students so they know exactly how to revise their writing. Reeves views writing as a way of thinking and believes that students process information in a much clearer way when they are required to write about it.

In *Writing Essentials,* Routman (2005) highlights the value of authentic writing. In conversations with teachers this is affirmed over and over again. Teachers consistently tell us that when students write about authentic experiences in science and social studies that the writing improves exponentially. The rocket story, in the Technological Design chapter, is a perfect example of this concept. If you recall, when the children wrote "How To Build a Rocket" papers based upon their own experiences the writing was simply terrific. The children had the authentic experience of building rockets and were able to transfer their personal experience to authentic writing.

Please note that the "Write to Inform" in science continuum below is based upon the Common Core Standards. The language from the standards was changed to "student friendly" reflective language and generalized across grade level bands. For more specificity please refer to the Common Core Standards.

Expository Writing Continuum Grades K-12 – Writing to Inform in Science (Student Language K-12)

This Continuum is based upon the Common Core Standards for English Language Arts and Literacy in History/Social Studies & Science

K-2	3-5	6-8	9-12
Write to Inform: I introduced my topic. I used facts and definitions to tell my reader more. I organized my writing under "big ideas". I ended with a sentence that brings my ideas together. I added information to a shared writing project using information from several books and my own experiences.	**Write to Inform:** a. I stated the topic clearly. b. I grouped related information in paragraphs or sections. c. I told more about the topic using facts, details, quotations and other information. d. I used linking words so my reader could easily understand the information I described. e. My conclusion was clear. It synthesized all of my ideas. **Gather Information:** • I gathered information from experience, print, and digital resources. • I summarized notes from multiple resources. • I provided basic bibliographic information.	**Write to Inform or Explain:** a. I provided an interesting introduction to my topic. b. I organized information under categories "big ideas". c. I used relevant facts, data, details and quotations to help my readers under each category/big idea. d. I used connecting vocabulary and clear sentence structure to create text that my reader could easily follow. e. I used scientific vocabulary accurately throughout my writing. f. I wrote objectively in a style that helps a reader seeking information. g. I wrote a conclusion that was clearly connected to the information. My conclusion synthesizes the information for my reader. **Gather Information:** I cited information effectively in a digital format. • I used multiple print and digital resources to gather information. • I carefully assessed the credibility and accuracy of each source before I used it. • I made certain that I quoted or paraphrased evidence, giving credit to the original source. • I respected the original author by avoiding plagiarism. • I followed an approved format for citation.	**Write to Inform or Explain:** a. I provided an interesting introduction to a complex topic. b. I organized the information so that each new piece of information built upon the piece that came before it to create a unified whole. c. I used formatting and graphics (e.g. headings, figures, tables, graphs and illustrations) to help my reader comprehend my writing. d. I developed a complex topic that is scientifically significant. e. I used data, quotations, and information from other individuals who have written about this topic. f. I used transitional words and phrases to help my reader understanding how the sections of my writing are connected. g. I used transitional words and phrases to link the different sections in my writing. h. My conclusion brings all the information together. It is a logical synthesis of the information. **Gather Information:** I cited information effectively in a digital format. • I used multiple print and digital resources to gather information. • I carefully assessed the credibility and accuracy of each source before I used it. • I integrated information from a variety of sources. • I avoided depending on one source as my major source of information. • I respected the original author. • I followed an approved format for citation.

Anne Grall Reichel, Ed.D.

FOR TEACHERS

Build Informational Text Rich Libraries In Your Classroom

Does your library have a balance of fiction and non-fiction? If not, then work towards that goal. Students seem to gravitate to non-fiction if it is available. Perhaps it is the intriguing photographs or logistical order of non-fiction that entices the curious mind. There seems to be a magnetic attraction between non-fiction and the curious child. The literacy experts have documented the fact that boys and struggling readers gravitate to non-fiction.

At the risk of supplying one story too many I am going to share an experience I had in a first grade classroom in Lake Bluff, IL. last fall. Mary Jo Stevenson, a first grade teacher, invited me to do a lesson with her first graders to bring closure to her insect unit. It was late October and we had already experienced an early frost in Chicago. I set out to find some interesting organisms to bring to Mary Jo's class. Despite the early frost the woods behind my home did not disappoint. Under leaf litter and logs I found plenty of diversity to build a living collection and more importantly to create an interesting inquiry that served as the catalyst to conversation with first graders. My collection included slugs, worms, spiders, insects and toads. (I assured the children that all creatures would be returned to the woods promptly after our lesson.) All creatures were carefully transported within leaf litter. When I arrived in Mary Jo's classroom there was great excitement surrounding the contents within the leaf litter. We set up stations around the classroom and challenged the children to support with evidence whether they thought each creature was or was not an insect. Before the young scientists set off to make their observations we operationally defined how we knew something was an insect. Our chart went something like this.

You know an organism is an insect if.....

- It has six legs
- Three body parts
- Antennae

The children were challenged to determine whether each of the ten organisms was an insect. They checked yes or no next to the number of the organism specimen on their recording sheet and were required to supply the evidence to support their determination. A sample appears below.

Organism #	Is the organism an insect? Write **yes** or **no**.	What is your evidence?
#1	No	The organism had eight legs. The organism only had 2 body parts.

After the children made their own personal decisions they returned to their table groups where further conversation took place. Before we knew it the table groups had managed to take books from the insect section of the classroom library to further their own discussion and scientific argumentation. Neither Mary Jo nor I had prompted the children to use books to support their determinations. The point being, if the books are available the children will use them even without being prompted to do so. All we have to do is supply the initial inquiry and the resources. I would like to add that my work takes me to both urban and suburban school. There is still great disparity between the classroom libraries I find in urban and suburban classrooms. We can do better than that for our children in urban settings!

A week after the lesson Mary Jo mailed me samples of expository writing. Mary Jo challenged the students to describe how to tell if something was an insect. They were directed to provide examples based upon their evidence. The writing was amazing. The first graders were capable of doing remarkable writing when provided with authentic experiences involving both reading and writing supported by inquiry-based science.

Link Scientific Drawing & Journal Entries to Writing

Labeled scientific drawings provide a great scaffold for writing. "The act of drawing and the picture itself both provide a supportive scaffolding." (Calkins, 1994, p.85) Digital images of children engaged in inquiry-based science can also be used. By supplying vocabulary and images children can readily make authentic connections, these connections in turn lead to authentic writing. Children use their drawings as a connection to the hands-on investigation and have easy access to the vocabulary because it appears on their drawings. The journal entries that often accompany scientific drawings do not require the formality involved in expository and persuasive writing. Journal entries often serve as the catalyst to spawn the formal writing process.

Look for the Myriad of Opportunities To Write to Inform in Science and Social Studies

One excellent scenario on writing in social studies comes from *How Students Learn* (2005). Please consult the book for a far more detailed explanation and a series of excellent questions. In the scenario children read several textbook accounts of the Pilgrim's arriving in America on the Mayflower. Next they look at an artistic portrayal and read William Bradford's journal entries. After great discussion, the children are challenged to evaluate the textbook accounts based upon reading the primary documents (Bradford's journal entries). Finally the children are challenged to write their own account. This kind of writing is authentic. The children become critical consumers of textbook information and have the opportunity to evaluate those entries based upon primary documents. This is far more authentic then the sit and read the textbook and respond to the questions scenario that is still alarmingly present in American classrooms.

As for science, expository writing is a natural. Children can write comparative pieces like, "How to determine if an insect goes through complete or incomplete metamorphosis". Applying technological design skills they can write about their authentic experiences on how to wire boxes, build rubber band powered cars or mouse trap cars. Applying systems thinking they can write about how to trouble-shoot a parallel or series circuit that is not functioning. They can write to inform on a topic of their choice. These can range from the impact of La Nina on Chicago winters to the likelihood of experiencing an earthquake in St. Louis. You get the idea. The last two examples require research skills and needles to say the application of reading strategies, note-taking skills and synthesis skills to mention a few.

In Barrington, every fourth grade student is involved in a field experience in which they collect prairie seed that is sown to restore additional prairie areas. We found that there were many natural opportunities for writing integration from expository to poetry. Here are a few examples from the Teacher Packet we developed with the assistance of fourth grade teachers in Barrington.

Barrington District 220: Integrating Writing With Our Prairie Preservation Field Experience

Expository	Why is the prairie important to us? or How to Preserve a Prairie
Persuasive:	Why preserve the prairie?
Narrative –	Day in the Life of a Bobolink, Shrew, or Herp
Poetry –	I am the Prairie – I am a Forb – I am a Grass

Barrington Teachers: Ron Bergquist, Theresa Boyd, Joslyn Katz, Marilyn Roselli, Kathleen Sanders, Terry Skibiski

---

Before leaving the topic of expository writing it is essential to address the dilemma middle school and high school teachers report regarding the time it takes to grade expository writing. Use the continuum to have students reflect on each component of the writing. Encourage peer review with the continuum as a guide. Provide formative feedback regularly based upon continuum expectations as students are writing. Invite revision. If you provide formative feedback do not take off points from the final grade. The spirit of formative feedback is to encourage students to reflect upon their work and to revise it. If we start taking off points each time we give feedback then savvy students are hard pressed to be motivated to improve their work.

FOR PARENTS

Build Informational Text Rich Environments In Your Home

In both the teacher section and the introduction, we have extolled the benefits of reading to improve writing. Take time to determine what your child is passionate about or nurture a passion by exposing your child to fascinating topics. From volcanoes, to earthquakes, to hurricanes, nature provides a wealth of invitations to inquiry. When you go to the library visit the non-fiction section with your child. Model your own interest as you open a few books. You modeling can go something like this. "I've always wanted to have time to figure out what causes tornadoes, look at this interesting book, I bet I can learn more about tornadoes if I check out this book and read it."

If you don't have time to go to the library go through the same explanation but use the world wide web. Model how you determine which sites are most reliable. For example, you might say something like this: "I think I will use this site on tornadoes because it is sponsored by NOAA, I know they are respected scientists." The point is our children need to see us involved in the pursuit of new information. Simply stated, they need to see us model our own fascination with learning something new and they need to see us model our own skepticism about the reliability of information readily available electronically. They need to hear us explain why we select a certain resource for information.

Consider the practice of giving Museum memberships or subscriptions to magazines like Time for Kids or National Geographic for Kids to the significant children in your life. These are gifts that promote reading in the content area. You simply can't fully appreciate a museum if you can't read and there is nothing better than having a child open the mailbox only to find their very own non-fiction magazine.

Use Science and Social Studies Topics to Converse

As mentioned earlier, conversation is a way of synthesizing ideas and provides a bridge from reading to writing. Margaret Wheatley reminds us that reflective conversation leads to the spawning of new ideas.

Think of all the times that you have solved a problem simply by talking it through with a good listener. Our children need us to be genuinely interested in what they have to tell us. Providing them with opportunities to become experts on topics of interest and subsequently listening to what they have to tell us helps children build great confidence.

Returning to the introductory thoughts at the beginning of the book. Somewhere along the way in America it became "geekish" to be well versed in a subject area. We need to bring back a sense of valuing the importance of learning. I remember one night a few years back when my husband and I were hosting a neighborhood dinner party. Two of the guests were out on the patio in a debate over the North Star that drew interest from several others. One neighbor shared his claim that the North Star remains in relatively the same position regardless of what time of night we observe it. The other neighbor thought this was incorrect and explained that like the other stars it appeared to move across the sky. My husband summoned me from the kitchen to resolve the debate. I responded that the North Star was circumpolar and that yes in fact it remained in relatively the same place all night and of course that was why it was a great navigation tool. As I spoke several people interested in the heated argument immediately scattered before my explanation was even complete. I of course stood there feeling like the "geek" almost wishing I had replied that I was not certain. When I look back on that evening I fully appreciate the way we make bookish kids feel. We silently give them the message that smart is abnormal and something to be hidden. We can do better than that by valuing learning and by showing interest in the subjects our children are passionate about. As I reflect on that evening I realize that we have become a nation who likes to argue. We point fingers left, right, upwards, downwards and sideways. We love the argument and ironically have little interest in the resolution. Once it is resolved there is nothing to sensationalize. The resolution is lost and we move on to the next drama. All too often our conclusions and claims are based on hearsay rather than on data, research and evidence.

Last year I visited my nephew who has two lovely, inquisitive young daughters. They had recently returned from the Grand Canyon and had fascinating questions about rocks, fossils, earth layers, etc.. When I returned home I put together a collection of books, fossils etc and shipped them off. My nephew reported that the girls immersed themselves in learning more for days. My advice is this, listen carefully and when a child asks about a rock or frog or storm, look at their question as the spark and as your invitation to support their inquiry. It is our job as adults to fuel the fire by responding with interest in the inquiry. The best way to ensure life long learning is by fueling the passion through books, links and tangible items to investigate. Have fun!

Write to Argue – (Supporting Claims With Evidence)

People in many walks of life have to support their claims with evidence. From people in sales and marketing to legal council you simply don't make the sale or convince the jury if you are incapable of providing well articulated, researched based evidence. Science and social studies are the perfect fodder to teach these skills.

The ICE Chart introduced in the chapter on Scientific Investigation is the perfect tool for supporting students in organizing their thinking on claims and evidence. The sample chart from the Scientific Investigation chapter is included below as a reminder.

Ideas My beginning ideas… What I think might happen	Claims What I found out	Evidence Data and/or observations that support my claim
All metals will be attracted to a magnet.	Not all metals are attracted to magnets.	The paper clip, nail, pin, scissors were attracted to the magnet. The penny, copper wire, and earring were not attracted to the magnet.

Students can use this chart when they begin a persuasive argument. It is important that they also understand that well developed persuasive arguments are proactive. By that we mean, the author of a persuasive argument is also fully aware of the counter argument and the evidence an individual on the other side of the argument might share.

Given the explosion of information that is readily accessible in a single swipe our children will need to be incredibly savvy in sorting out the many claims on any given subject. They will need to be able to process a great deal of information before arriving at an intelligent conclusion.

The "Write to Argue" continuum, like the "Write to Inform" continuum, was informed by the Common Core Standards. As mentioned previously the salient points were reworded in kid friendly language in grade level bands. For more specificity please refer to the Common Core Standards. A link is provided in the reference section in the back of the book.

Write to Argue
Continuum Grades K-12 – Writing in Science to Support Claims With Evidence

Please Note: This continuum was developed from the Common Core State Standards. In some places there is a blending of the Common Core Standards with the National Science Education Standards and Benchmarks for Science Literacy. Some modifications have been made to reflect the work of the NSTA Anchors Project. Specifically the language "claims" and "evidence" reflects the thinking processes explicated in The Nature of Science section of the Anchors Project.

<u>Observation</u>: Writing standards embed the "Nature of Science" and writing tasks require the ability to synthesize many of the skills related to "Habits of Mind". The ability to articulate and communicate a scientific "claim" supported with "evidence" is central to the work of the scientist in the field.

K-2	3-5	6-8	9-12
I stated a "claim". I supported my "claim" with "evidence" from my science journal, talks with other scientists in my class, and from the books I read. I used words like (e.g. *because* and *also*) when connecting my claims with evidence. Please note CCSS uses the terms *opinion* and *detail*. "Claim" and "Evidence" were supplemented to represent the thinking process explicated in **The Nature of Science.** With guidance young children can use this language operatively.	I stated a "claim" and supported my "claim" with "evidence". a. I stated my "claim". b. I created an organized way of sharing my "claim" and "evidence" with my reader. c. I provided "evidence" from my data, conversations with other scientists, and my research. d. I linked my ideas together with words. (e.g. *consequently, generally, specifically*) e. I kept my reader interested and used persuasive language in my writing. f. I organized my thinking in a concluding sentence that restates my main idea. Please note CCSS uses the terms *opinion* and *detail*. "Claim" and "Evidence" were supplemented to represent the thinking process explicated in **The Nature of Science.**	I wrote an argument focused on science content. a. I introduced a claim about a topic or issue. b. I distinguished my claim from alternate or opposing claims. c. I organized reasons, data, and evidence logically to support my claim. d. I supported the claim with logical reasoning and detailed, accurate data and evidence. e. I used words and phrases as well as science vocabulary to make clear the relationships among claims, reasons, data, and evidence. f. I used an objective style and tone throughout my writing. g. I concluded my writing with a statement that follows logically from my argument.	I wrote an argument focused on science content. a. I introduced a claim about a topic or issue. b. I distinguished my claim from alternate or opposing claims. c. I organized reasons, data, and evidence logically to support the claim. d. I developed my claim thoroughly supplying the most relevant data and evidence acquired in a scientifically acceptable form. e. I used scientific vocabulary and precise words to clarify the relationships between claims and evidence. f. I maintained an objective style and tone throughout my writing. g. I concluded my writing with a statement that follows logically from my argument.

FOR TEACHERS

The continuum ideas set forth throughout the book apply here as well. The important thing is to supply interesting topics that will engage children in the process of supporting claims with evidence. As mentioned earlier the more authentic the task the greater the engagement. This can start in the early grades and continue to grow as students move through their schooling. A few ideas follow.

Primary children often do a unit on solids and liquids in science. We can raise the bar on the level of complexity of student investigations by moving beyond simple classification of solids and liquids. We can get into the heat energy interaction in very concrete ways. Students can be challenged to conduct fair tests to determine which insulating materials work best to keep ice frozen. After the students conduct fair tests and collect data they can process their data on tables and then be challenged to support with evidence which insulating materials they would choose to build a popsicle saving device. Next the children can be challenged to state which insulating materials they recommend that you should purchase to build popsicle savers. Insist that they supply the evidence from their hands on investigations. This kind of thinking in science leads directly into persuasive writing. Don't stop here. Purchase the materials based upon the recommendations and build the popsicle savers. It works! First grade teachers in Barrington do this project.

Children in grades 3–5 are capable of remarkable levels of argumentation when provided with the appropriate scaffold. Third graders in Barrington compare models to real objects. We start with comparing toy trucks to real trucks. Next the children participate in a scale modeling project of the solar system with the goal of helping children construct an understanding of just how vast the solar system is. Finally we challenge them to examine illustrations of the solar systems in a variety of informational text. We then challenge the students to write back to the illustrators, supporting with evidence from their experience just why the illustrator may want to reconsider their illustration. For example, the children discover many images where Jupiter is larger than the sun or where earth appears much larger than Venus. Because the writing is based upon concrete experiences with scale modeling and critical examination of illustrations the children readily support their claims to the illustrators with evidence.

Students in grades 6–12 are ready to engage in conversations about important issues from debating measures to remove the Asian Carp from the Chicago River to the funding of Space Exploration. They hear and read about all kinds of medical advances and listen to the risks and benefits of certain interventions. There isn't a skeptic amongst us who hasn't listened to some of the pharmaceutical commercials and considered the notion that the ailment might be better than the side effects of the drug. The questions, available research and opportunity to develop positions are limitless. As teachers we must be vigilant in communicating the importance of checking the reliability of sources. We must encourage students to seek evidence from multiple sources and we must encourage them to embrace opposing views to better inform their position.

FOR PARENTS

The recommendations are rather simplistic and very brief. We need to model our thinking. When we state our position on something it is important to explain the evidence behind our position. When our children make any kind of claim we should challenge them to support their thinking with evidence.

It is interesting to share editorials from the newspaper or letters to the editor and talk about them. Don't ask your child to side with one or the other, just ask them to tell you which writer best supports their position. Challenge them to explain why.

IN CONCLUSION

Please consider this work to be an invitation to **expect more.** Find time to engage the children in your world in meaningful conversations and meaningful work. It would be difficult to argue that we are living in challenging times. Our children will need the skills of resiliency and perseverance. They will need to understand that growth and authentic learning come only after careful reflection and the willingness to improve upon their initial work.

I recognize that this work will benefit greatly from many more examples of each continuum in action and from samples of student work generated after reflective use of the continuums. I invite you to use the ideas and share the ways in which you modify and improve upon the work herein. I consider this work to be a starting point.

In *Pale Blue Dot* Carl Sagan argued that the American spirit was founded on the interest in exploring new frontiers. Perhaps our new frontier could be our collective interest in reigniting the love of learning. Imagine if each reader took the time to discover what sparks the interest of a child they know and then went on to support that child's interest by learning along with them. Just imagine!

Appendix A: Why Should We Use a Unifying Concept?

Some important themes pervade science, mathematics and technology
and appear over and over and over again, whether
we are looking at an ancient civilization,
the human body, or a comet.
They are ideas that transcend disciplinary boundaries
and prove fruitful in explanation,
in theory, in observation and in design.
SCIENCE FOR ALL AMERICANS

- As reported in *How People Learn,* knowing facts is not enough; facts and ideas become usable knowledge when learners connect and organize them in meaningful ways around unifying concepts.

- The unifying concepts build upon student understandings learned in previous years.

- The unifying concepts afford students and teachers with the opportunity to make connections in and amongst the disciplines of life, physical and earth/space science.

- The unifying concepts focus on depth of understanding rather than breadth.

- The unifying concepts play an integral role in authentic integration of curriculum.

SAMPLE VERTICAL USE OF THE UNIFYING CONCEPT

Kindergarten: Systems of Sorting	Life	Physical	Earth
	Sort organisms by observable characteristics including coverings, # of legs, movement, type of teeth, etc.,	Sort objects by observable properties: color, texture, shape, size, sink/float, magnetic/non-magnetic	Sort earth materials by observable properties: shiny/dull, rough/ smooth, etc.,

First Grade: Change Over Time	Life	Physical	Earth
	Habitats change over time: # of plants, # of animals, time of year	Solids, liquids and gases change over time	Weather changes over time
Make connections to Systems of Sorting	Group animals as mammals, reptiles, birds, etc., Group animals by where they live, Group nocturnal and diurnal animals	Sort matter as solids, liquids and gases, sort changes as fast and slow,	Sort clouds, sort rainy days, sunny days, etc.,

Second Grade: Patterns of Change	Life	Physical	Earth
	Insect Life Cycles: Compare and contrast patterns of change in life cycles of insects	Sound: There is a pattern when we look at the following – the higher the pitch, the thinner the rubber band, the lower the pitch the thicker the rubber band	Sun, Moon & Earth The length of a shadow changes throughout the day, the pattern is predictable. The earth has a day/night cycle (pattern), The moon phases are predictable and provide evidence of a pattern of change.
Make connections to Systems of Sorting	Group insects that go through complete and incomplete metamorphosis	Classify pitches as high or low, volume as loud or soft	Classify shadows as long or short, classify sky objects associated with night and day
Make connections to Change Over Time	Insects change over time	We can change the pitch and volume of a sound. Light changes (refracts/bends) as it moves through different medium	The position of the sun appears to change throughout the day. The moon changes throughout the month.

Third Grade:	Life	Physical	Earth
Systems & Relationships	A plant is a system. Plants are dependent upon relationships with pollinators in order to produce new seeds. A prairie ecosystem is a system There are relationships in and amongst organisms in a prairie ecosystem.	Force & Motion: Simple Machines combined together create systems. There is a "trade off" (relationship between force and distance when we use a simple machine.	Astronomy: We are part of a solar system. There is a relationship between the pull of gravity and the mass of an object. There is a relationship between the position of the earth and sun system and the season we experience.
Make connections to <u>Systems of Sorting</u>	We can group plant parts (seeds, roots, stems, flowers). We can group prairie organisms is producers, consumers and decomposers	We can group simple machines by their function.	We can group planets by their characteristic properties (Example rocky and gaseous)
Make connections to <u>Change Over Time</u>	Plants change over time, prairie systems change over time.	We can change the amount of force needed to move an object by changing the position of a simple machine.	The phase of the moon we see changes as a result of our relationship with the moon in terms of relative position.
Make connections to <u>Patterns of Change</u>	The plant life cycle is a pattern of change. We can predict changes in prairie populations by observing patterns. (Example: If a population of predators increases in size we can predict a decrease in the population of prey.)	If we graph force and distance when using a simple machine we will be able to infer a pattern. Example: The height of an inclined plane to the amount of force required to move an object.	Rotation and revolution of planets are predictable patterns of change

Fourth Grade:	Life	Physical	Earth
Change & Constancy "Things change in steady, repetitive, or irregular ways – or sometimes in more than one way at the same time." AAAS	Ecosystems are dynamic systems that can change in steady, repetitive and irregular ways.	Electricity & Magnetism Electric circuits are systems that can change as a result of interactions including position of a battery, insertion of conductors and insulators, etc., Magnetic forces can change as a result of relative position, example north to north pole	Land & Water Water cycles (changes). Land changes over time as a result of forces such as ice, water, wind, rain. Some changes to land and water can be steady while others are irregular.
Make connections to Systems of Sorting	Sort predators/prey, consumers, producers and decomposers, diverse ecosystems, adaptations associated with diverse systems.	Group conductors and insulators, objects that are magnetic and non-magnetic, parallel and series circuits	Group changes to land as slow and fast, constructive and destructive, group clouds (as studying water cycle)
Make connections to Change Over Time	Diverse ecosystems change over time. Adaptations change over time.	We can change a circuit by adding additional batteries, by opening or closing a switch, etc.,	Water changes as it cycles from solid, to liquid to gas. Land changes as a result of interactions with water causing weathering and erosion
Make connections to Patterns of Change	There are patterns of change associated with diverse ecosystems. Some patterns within ecosystems are predictable. (Example increase in one population may cause a decrease in another population)	There are predictable patterns when working with circuits. Example: If I increase the number of batteries I will increase the brightness of a bulb.	There are patterns in changes in Land & Water. Example: If I increase the flow of water then I increase the amount of deposition down stream.
Make connections to Systems & Relationships	Each diverse ecosystem is a system with many relationships.	Circuits are systems. There is a relationship between a change in one part of a circuit to another.	A river is a dynamic system. The interactions of land and water are relationships.

Fifth Grade:	Life	Physical	Earth
Systems & Interactions	The Human Body is a dynamic system. Interactions: Diet, Exercise, Bacteria, Viruses, Drugs, etc.,	Motion & Design Our vehicles are systems. Our vehicles interaction with gravity, friction, air resistance, etc.,	Impact Earth Microworlds are tiny systems they are impacted by interactions. Interactions on a microscopic level can have a macroscopic impact.
Make connections to Systems of Sorting	Sort cells, tissues, organs, organ systems.	Group balanced and unbalanced forces, potential and kinetic energy, actions and reactions	Group actions as having a positive and negative impact on the environment. Group microscopic and macroscopic organisms.
Make connections to Change Over Time	Body systems change over time.	Forces change as a result of interactions	Our climate appears to be changing over time.
Make connections to Patterns of Change	There are patterns of change associated with choices we make in caring for our fragile human systems.	There are many patterns associated with motion and design. Example: If I increase the mass of a vehicle then I will need a greater force to move an object.	There are patterns in microscopic communities. If we change the acidity in solution we can expect a change in population size of vinegar eels. Climate patterns have been studied over years, they appear to be changing.
Make connections to Systems & Relationships	Human Body is a system. There are relationships in and amongst body systems.	Our vehicle is a system. Example: There is a relationship between force and distance traveled.	Our earth is a dynamic system. Our relationship with the earth impacts the overall health of the planet from a microscopic to a macroscopic level.
Make connections to Change & Constancy "Things change in steady, repetitive, or irregular ways – or sometimes in more than one way at the same time." AAAS	Human Body systems change in regular and irregular ways.	As we work with vehicles there are many regular patterns of change associated with the ways in which we change vehicles. Example: Wheels with and without tires.	Our planet seems to be changing in regular and irregular ways.

Appendix B: Citizen Scientist & Leave No Child Inside Links

Cornell Lab of Ornithology
PROJECTS
- Feeder Watch
- Pigeon Watch
- Nest Watch

http://www.birds.cornell.edu/pfw/

Create a Certified Wildlife Habitat
Provides information for creating a wildlife habitat in your own backyard
http://www.nwf.org/gardenforwildlife/create.cfm?CFID=22006873&CFTOKEN=38c50
7d0d35e6473-B5405AE2-5056-A868-A00493F00E3AD0DE

EarthTrek projects involve people collecting data outside using their GPS unit to locate a feature, or to record the coordinates of a data point.
http://www.goearthtrek.com/

Great Backyard Bird Count engages the public in systematically counting the birds in their own backyards
http://www.birdsource.org/gbbc/

Journey North
A global study of wildlife migration and seasonal change
http://www.learner.org/jnorth/

Leave No Child Inside Consortium
Ideas to get kids outdoors
http://funoutside.org/

Last Child in the Woods
http://richardlouv.com/

Project BudBurst engages the public in making careful observations of the first leafing, first flowering and first ripening of fruits.
http://www.windows.ucar.edu/citizen_science/budburst/

The Lost Ladybug Project
Photograph the ladybugs you find
http://www.lostladybug.org/

Appendix C: Technological Design Links

Kids Design the Future
http://www.cs.umd.edu/hcil/kiddesign/

Children's Engineering
Developing technological literacy at the elementary level
http://www.childrensengineering.com/

The Design Squad Educators Guide
http://pbskids.org/designsquad/parentseducators/educators_guide.html

Engineering for Kids
http://www.biglearning.com/treasureengineering.htm

Children's Engineering
http://www.childrensengineering.com/teacherresources.htm

Zoom Projects for Kids
http://pbskids.org/zoom/activities/sci/

Design Technology Engineering Projects
http://www.ncsu.edu/kenanfellows/?q=node/199

How Stuff Works
http://www.howstuffworks.com

Appendix D: Children's Books About Scientists

Cherry, L., & Braasch,G. 2008. *How we know what we know about our changing climate: Scientists and kids explore global warming.* Nevada City, CA: Dawn Publications.

Harper, Charice. 2001. *Imaginative inventions.* New York, NY: Little Brown & Company.

Hakim, Joy. 2004. *The story of science: Aristotle leads the way.* Washington & New York: Smithsonian Books.

Hakim, Joy. 2005. *The story of science: Newton at the center.* Washington & New York: Smithsonian Books.

Hakim, Joy. 2007. *The story of science: Einstein adds a new dimension.* Washington & New York: Smithsonian Books.

Jackson, Donna. 2009. *Extreme scientists: Exploring nature's mysteries from perilous places.* New York, NY: Houghton Mifflin Harcourt.

Locker, Thomas. 2003. *John Muir america's naturalist.* Golden, CO: Fulcrum Publishing.

Locker, T., & Bruchac, J. 2004. *Rachel Carson: Preserving A sense of wonder.* Golden, CO: Fulcrum Publishing.

Martin, Jacqueline. 1998. *Snowflake Bentley.* Boston, MA: Houghton Mifflin.

Montgomery, Sy. 2004. *The tarantula scientist: Scientist in the field series.* Boston, MA: Houghton Mifflin.

Panchyk, Richard. 2005. *Galileo for kids.* Chicago, IL: Chicago Review Press.

Pettenati, Jeanne. 2006. *Galileo's journal.* Watertown, MA: Charlesbridge Publishing.

Sis, Peter. 1996. *Galileo Galilei.* New York, NY: Frances Foster Books.

References

American Association for the Advancement of Science. (1993). *Benchmarks for Science Literacy.* New York, NY: Oxford University Press.

American Association for the Advancement of Science. (1989). *Science for all americans.* New York, NY: Oxford University Press.

Bransford, J., Brown, A., & Cocking, R., eds. (2000). *How People Learn: Brain, Mind, Experience and School.* Washington, D.C. National Academy Press.

Calkins, L. (1994). *The art of teaching writing.* Portsmouth, NH: Heinemann.

Campbell Hill, B. (2001). *Developmental Continuums: A Framework for Literacy Instruction and Assessment K-8.* Norwood, MA: Christopher-Gordon Publishers.

Donovan, M., & Bransford, J., eds. (2005). *How Students Learn History, Mathematics, and Science in the Classroom.* Washington, D.C. The National Academies Press.

Dueschl, R., Schweingruber, H., & Shouse, A. eds. (2007). *Taking science to school: Learning and teaching science in grades k-8.* Washington, D.C. The National Academies Press.

Hakim, Joy. 2007. *The Story of science: Einstein adds a new dimension.* Washington & New York: Smithsonian Books.

Hyde, A. (2006). *Comprehending math: Adapting reading strategies to teach mathematics, K-6.* Portsmouth, NH: Heinemann.

Jackson, Donna. 2009. *Extreme scientists: Exploring nature's mysteries from perilous places.* New York, NY: Houghton Mifflin Harcourt.

Keeley,P., Eberle,F., & Farrin,L. (2005). *Uncovering student ideas in science: 25 formative assessment probes.* Arlington, VA: NSTA Press.

Keeley,P., Eberle,F., & Tugel,J. (2007). *Uncovering student ideas in science: 25 more formative assessment probes.* Arlington, VA: NSTA Press.

Keeley,P., Eberle,F., & Dorsey,C. (2008). *Uncovering student ideas in science: Another 25 formative assessment probes.* Arlington, VA: NSTA Press.

Keeley,P., & Tugel,J. (2009). *Uncovering student ideas in science: 25 new formative assessment probes.* Arlington, VA: NSTA Press.

Keene, E., & Zimmermann, S.(1997). *Mosaic of thought: Teaching comprehension in a reader's workshop.* Portsmouth, NH. Heinemann.

Lemke, J. (2004). "The literacies of science." In *Crossing borders in literacy and science instruction.* Arlington, VA. NSTA Press & Newark, DE: International Reading Association

Louv, R. (2005) *Last child in the woods: Saving our children from nature- deficit disorder.* New York, NY: Algonquin Books

Marzano,R., Pickering,D., & Pollock,J. (2001). *Classroom instruction that works: Research-based strategies for increasing student achievement.* Alexandria, VA: ASCD.

Mathewson, JH (1999). Visual–spatial thinking: an aspect of science overlooked by educators. Science Education, 83, 33–54.

Miller, D. (2002). *Reading with meaning: Teaching comprehension in the primary classroom.* Portland, ME: Stenhouse.

National Academy of Sciences. (2002). *Science and technology for children.* Burlington, NC: Carolina Biological Supply Company.

Osborne, R., & Freyberg, P. (1985). *Learning in science. The implications of children's science.* Portsmouth, MA: Heinemann Education.

Pearson,P. & Cervetti, G. (2005). Reading and writing in the service of acquiring scientific knowledge and dispositions: In search of synergies. A presentation given at the International Reading Association Annual Research Convention, San Antonio, TX, April 30.

Pink, Daniel. (2005) *A whole new mind.* New York, NY: Riverhead Books.

Posner, G. J., Strike, K. A., Hewson, P. W., & Gertzog, W. A. (1982). Accommodation of a scientific conception: Towards a theory of conceptual change. *Science Education,* 66 (2), 211-227.

Reeves, D. (2002). *The daily disciplines of leadership.* San Francisco, CA: Jossey–Bass.

Routman, R. (2005). *Writing essentials: Raising expectations and results why simplifying teaching.* Portsmouth, NH: Heinemann.

Sahn, L. & Reichel, A. (2008). "Read all about it! A classroom newspaper integrates the curriculum. *Young Children*. 63 (2): 12-18.

Stotsky, S. (1983). "Research on reading/writing relationships: A synthesis and suggested suggestions." *Language Arts* 60: 627-42.

Van Dyke, F. (1998). Visual Approaches to Algebra. Dale Seymour Publication.

Wheatley, M. (1999). *Leadership and the new science: Discoverin order in a chaotic world*. San Francisco, CA: Berrett-Koehler Publishers.

Index